AF781192

21st Century Advanced Carbon Materials for Engineering Applications - A Comprehensive Handbook

Edited by Mujtaba Ikram and Asghari Maqsood

Published in London, United Kingdom

IntechOpen

Supporting open minds since 2005

21st Century Advanced Carbon Materials for Engineering Applications - A Comprehensive Handbook
http://dx.doi.org/10.5772/intechopen.92494
Edited by Mujtaba Ikram and Asghari Maqsood

Contributors
Mujtaba Ikram, Sadia Sharif, Sana Arbab, Amna Saeed, Khurram Shahzad, Muhammad Aamir Iqbal, Abdullah Khan Durrani, Asghari Maqsood, Muhammad Ikram, Ali Raza, Sarfraz Ali, Salamat Ali, Atif Shahbaz, Haleema Ijaz, Ali Haider, M Tayyab Hussain, Junaid Haider, Arsalan Ahmed Rafi, Khalid Parwez, Asim Hassan Rizvi, Aneeqa Bashir, Azka Mehvish, Maria Khalil, Arun A. Bhagwath, Asif Zawed, Bhagwan Rekadwad, Suman V. Budihal

© The Editor(s) and the Author(s) 2021
The rights of the editor(s) and the author(s) have been asserted in accordance with the Copyright, Designs and Patents Act 1988. All rights to the book as a whole are reserved by INTECHOPEN LIMITED. The book as a whole (compilation) cannot be reproduced, distributed or used for commercial or non-commercial purposes without INTECHOPEN LIMITED's written permission. Enquiries concerning the use of the book should be directed to INTECHOPEN LIMITED rights and permissions department (permissions@intechopen.com).
Violations are liable to prosecution under the governing Copyright Law.

CC BY

Individual chapters of this publication are distributed under the terms of the Creative Commons Attribution 3.0 Unported License which permits commercial use, distribution and reproduction of the individual chapters, provided the original author(s) and source publication are appropriately acknowledged. If so indicated, certain images may not be included under the Creative Commons license. In such cases users will need to obtain permission from the license holder to reproduce the material. More details and guidelines concerning content reuse and adaptation can be found at http://www.intechopen.com/copyright-policy.html.

Notice
Statements and opinions expressed in the chapters are these of the individual contributors and not necessarily those of the editors or publisher. No responsibility is accepted for the accuracy of information contained in the published chapters. The publisher assumes no responsibility for any damage or injury to persons or property arising out of the use of any materials, instructions, methods or ideas contained in the book.

First published in London, United Kingdom, 2021 by IntechOpen
IntechOpen is the global imprint of INTECHOPEN LIMITED, registered in England and Wales, registration number: 11086078, 5 Princes Gate Court, London, SW7 2QJ, United Kingdom
Printed in Croatia

British Library Cataloguing-in-Publication Data
A catalogue record for this book is available from the British Library

Additional hard and PDF copies can be obtained from orders@intechopen.com

21st Century Advanced Carbon Materials for Engineering Applications - A Comprehensive Handbook
Edited by Mujtaba Ikram and Asghari Maqsood
p. cm.
Print ISBN 978-1-78985-912-6
Online ISBN 978-1-78985-924-9
eBook (PDF) ISBN 978-1-78985-991-1

For EU product safety concerns:
IN TECH d.o.o., Prolaz Marije Krucifikse Kozulić 3, 51000 Rijeka, Croatia,
info@intechopen.com or visit our website at intechopen.com.

We are IntechOpen, the world's leading publisher of Open Access books

Built by scientists, for scientists

5,500+
Open access books available

135,000+
International authors and editors

165M+
Downloads

156
Countries delivered to

Our authors are among the
Top 1%
most cited scientists

12.2%
Contributors from top 500 universities

WEB OF SCIENCE™

Selection of our books indexed in the Book Citation Index
in Web of Science™ Core Collection (BKCI)

Interested in publishing with us?
Contact book.department@intechopen.com

Numbers displayed above are based on latest data collected.
For more information visit www.intechopen.com

Meet the editors

Dr. Mujtaba Ikram is an assistant professor at the University of the Punjab, Lahore, Pakistan. He is also a visiting scholar at the Abdus Salam International Centre for Theoretical Physics (ICTP), Italy. His research interests include nanotechnology, renewable energy, material science, and engineering. His work has been cited by scientists from all over the world. He has several research publications, book chapters, reviews, and proceedings to his credit. He has represented his research in the United States, Italy, Egypt, Germany, Slovenia, China, Hong Kong, Malaysia, United Arab Emirates, and many other countries. Dr. Ikram was selected among two young scientists from South Asia for The World Academy of Sciences (TWAS) science diplomacy held in Trieste, Italy, in 2013. He has been invited as a lecturer many times by the Chinese Academy of Sciences (CAS)-TWAS Beijing. In 2015, he was awarded the CAS-TWAS green technology award. In 2017, he was awarded the CAS-TWAS green chemistry and technology (GCT) award for his guest lectures. He has also been awarded various prestigious fellowships throughout his academic and professional career.

Professor Emeritus Dr. Asghari Maqsood is currently working as an advisor to the vice-chancellor, Air University, Islamabad, Pakistan, where she also served as a dean in the Faculty of Basics and Applied Sciences. She has more than forty-eight years of experience in the research of advanced materials. She obtained her MSc from Oxford University, and Ph.D. in Materials Science from Goteborg University, Sweden, along with a diploma from Uppsala University, Sweden. She has 212 research publications to her credit including 180 journal publications and 4 book chapters. She has been awarded many national and international awards including a Gold Medal from the Pakistan Academy of Sciences (2000), President's Award for Pride of Performance (2001), HEC Best University Teacher Award (2002), Prime Minister Gold Medal (2004), Izaz-i-Fazeelat for Academic Distinction (2005), and Civil Award Sitara- e- Imtiaz (2010). Recently, her name appeared among the world's top 2% of scientists on a list by Stanford University, California, USA.

Contents

Preface

Advanced carbon materials represent the next scientific revolution, especially in the fields of nanotechnology, drug delivery, and materials sciences. Advanced carbon materials such as graphene, fullerenes, and carbon nanotubes (CNTs) are heavily researched nanostructures. Due to their outstanding physical properties, advanced carbon monoliths have been used in photovoltaic, environmental, energy, thermal, and electronic applications. The 21st century is considered the "scientific era of graphene," which is a form of carbon with desirable properties such as high electrical conductivity, thermal conductivity, strength, and permeability. Graphene is a material that conducts electricity and heat maximally, making it an ideal candidate for energy and thermal applications, respectively. The book includes comprehensive information on fabrication, emerging physical properties, and technological applications of advanced carbon materials.

The book is organized into three sections, each of which contains two detailed and comprehensive chapters. Section 1, "Advanced Carbon Materials: A Comprehensive Overview," includes Chapter 1, "Introductory Chapter: Introduction to Advanced Carbon Materials and Innovative Engineering Applications, and Chapter 2, "Advanced Carbon Materials: Base of 21st Century Scientific Innovations in Chemical, Polymer, Sensing and Energy Engineering."

Chapter 1 provides detailed insight into various forms of advanced carbon materials, including graphene, structural graphite, CNTs, diamond-like carbon (DLC), carbon foam, and fullerene. It discusses both physical properties and applications of advanced carbon materials. Chapter 2 reviews the conjugation of graphene with other 2D materials to produce further remarkable compounds appropriate for an extensive variety of applications. The chapter examines the uses and applications of advanced carbon materials in chemical, polymer, sensing, and energy engineering.

Section 2, "Sustainable Fabrication of Advanced Carbon Materials for Applied Applications," includes Chapter 3, "Carbon Nanotubes Integrated Hydroxyapatite Nano-Composite for Orthopaedic and Tissue Engineering Applications" and Chapter 4, "Advanced Carbon Materials for Sustainable and Emerging Applications."

Chapter 3 provides fabrication details for CNTs with significant physical properties for applied applications in orthopaedics and tissue engineering. Chapter 4 describes advanced carbon materials for sustainable applications in electronics, energy conservation, drug delivery, and more. These advanced materials are the future of sustainable energy production and storage devices owing to their capability to store energy on a large scale. In the near future, fuel cells will likely replace battery-based energy systems, and graphene sheets will be game-changers in microelectronics.

Section 3, "Electrochemistry of Advanced 2D Carbon Materials," includes Chapter 5, "Electrochemical Exfoliation of 2D Advanced Carbon Derivatives" and Chapter 6, "Carbon Nanotubes."

"Chapter 1, Introductory Chapter: Introduction to Advanced Carbon Materials and Innovative Engineering Applications" will have a detailed insight into various forms of advanced carbon materials especially graphene, structural graphite, carbon nanotubes, diamond-like carbon (DLC), carbon foam and fullerene, respectively. The chapter grants detailed insight into physical properties and applications of advanced carbons materials.

"Chapter 2: Advanced Carbon Materials: Base of 21st Century Scientific Innovations in Chemical, Polymer, Sensing and Energy Engineering" will have a detailed insight with respect to conjugation of graphene with other 2D material which will be developed to produce further remarkable compounds that make it appropriate for an extensive variety of applications. This chapter grants the utilization and applications of advanced carbons materials in chemical, polymer, sensing and energy engineering.

"Chapter 3: Carbon Nanotubes Integrated Hydroxyapatite Nano-Composite for Orthopaedic and Tissue Engineering Applications" will give fabrication details for Carbon nanotube with significant physical properties for applied applications in orthopaedics and tissue engineering.

"Chapter 4: Advanced Carbon Materials for Sustainable and Emerging Applications" will describe in-depth details of advanced carbon materials as a sustainable prospectus. Advance carbon nanomaterials are receiving a lot of attention from scientific research in the last few years owing to their unique mechanical, thermal, chemical, optical and electrical properties. Advanced carbon nanomaterials, comprising of graphene, fullerene, carbon fibers, activated carbon and carbon nanotubes are considered as the backbone of material science and technological innovation. These nanomaterials are fabricated by using different physical and chemical methods to get high materials with excellent characteristics. Advanced carbon materials also find applications in electronics, organic photovoltaic, energy conservation technology and drug delivery, etc. In the future, these advanced materials can be used to develop several materials with different applications. A lot of research is taking place for producing these materials on the industrial level. These advanced materials are the future of sustainable energy production and storage devices owing to their capability to store energy on large scale. Fuel cells also in the near future are thought to replace battery-based energy systems. Graphene sheets may have the potential to be a game-changing use in microelectronics. The demand for advanced carbon materials will continue to grow for technological innovation.

Chapter 5 talks about advanced 2D carbon materials such as graphene and its derivatives, which are basic building blocks for future nanostructures in electronics and energy owing to their remarkable physical and chemical properties. A variety of techniques have been employed to develop 2D advanced carbon materials, including state-of-the-art synthetic protocols like electrochemical exfoliation, which provides high yield, great performance, low cost, and excellent scalability. Notably, playing with electrochemical parameters not only allows tuneable properties but also enhances the content variety from graphene to a wide spectrum of 2D semiconductors. This chapter provides a succinct and comprehensive survey of recent progress in electrochemical exfoliation routes and presents the processing techniques, strategic design for exfoliations, mechanisms, and electrochemistry of graphene.

Chapter 6 describes CNTs, also referred to as carbon nano-architecture allotropes, with wrapped graphene sheets forming a cylindrical structure. CNTs are either developed by metals or narrow-band semiconductors with rolling graphene sheets in various ways. CNTs possess certain peculiar properties, such as a high degree of stiffness, wide ratio of length to diameter, and remarkable toughness, and are employed in several applications. These properties can be enhanced by scheming the diameter, nature of walls, chirality, and length depending on the synthesis process. This chapter examines the various properties of CNTs and explains numerous methods of synthesis and processing of CNTs, comparing advantages and drawbacks.

I acknowledge the Higher Education Commission Start Up Research Grant (2410) for financial support. I am deeply thankful to my coeditor, Professor Emeritus Dr. Asghari Maqsood, for expert advice and critical comments. I am thankful for the support of my family, without which this book would not have been possible.

Dr. Mujtaba Ikram
Institute of Chemical Engineering and Technology (ICET),
University of the Punjab,
Lahore, Pakistan

Dr. Asghari Maqsood
Professor,
Department of Physics,
Air University,
Islamabad, Pakistan

Section 1

Advanced Carbon Materials: A Comprehensive Overview

Chapter 1

Introductory Chapter: Introduction to Advanced Carbon Materials and Innovative Engineering Applications

Sadia Sharif, Sana Arbab, Amna Saeed, Khurram Shahzad, Muhammad Aamir Iqbal, Abdullah Khan Durrani, Asghari Maqsood and Mujtaba Ikram

1. Introduction

Carbon belongs to the group IV of periodic table with atomic number 6. Graphite, diamond and fullerene are the allotropic forms of carbon. Carbon-based materials are versatile in term of applications due to its nature to chemically combine with other carbon-based materials and to make a strong covalent bond with a range of different elements. Therefore, they have outstanding properties like high strength, high density, and high hardness. The amazing characteristics of carbon materials make them most suitable candidate in various applications of advance technology. Graphene, carbon fibers, carbon foams, structural graphite (special graphite), carbon nanotubes (CNT), diamond-like carbon (dlc) and nano-crystalline diamond (ncd) are all included in advance carbon materials category. Advance carbon material are backbone of next generation scientific revolution especially in field of nanotechnology and materials sciences, respectively. Advanced carbon materials which are graphene, fullerenes, CNTs are considered most researched nanostructures in last couple of decades. Due to outstanding physical properties of advanced carbon monoliths, it has been used in photovolatic, environment, energy, thermal, and electronic applications. 21st century is being considered as "Scientific era of graphene", which is most amazing form of carbon due to highest electrical conductivity, thermal conductivity, strength and permeabile properties. Graphene is a material which conducts electricity and heat to maximum, which makes it an ideal candidate for energy and thermal applications, respectively. The following chapter will have a detailed insight in various forms of advanced carbon materials especially graphene, structural graphite, carbon nanotubes, diamond-like carbon (DLC), carbon foam and fullerene, respectively. The chapter grants detailed insight into physical properties and applications of advanced carbons materials.

2. Types of advanced carbon materials and related application

2.1 Graphene

Graphene is first ever 2-D allotropic form of carbon with hybridized Sp2 bonding which gave rise to new advancements in research and technology. The mysterious

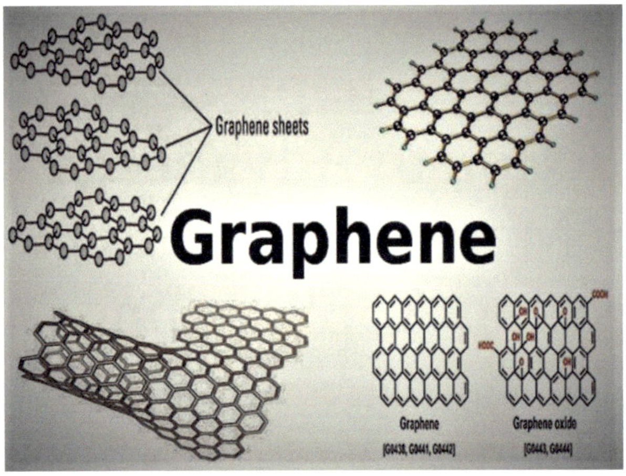

Figure 1.
Descriptive illustration of various structures of graphene.

electronics, structural, electrochemical, and physicochemical features of single layer of this material are significantly the big concern for the material scientists [1–4]. Mono atomic carbon layer extending across the two provided large surface areas due to an incredibly porous material [5], representing its potent absorbing ability. Hereby, scientists may conclude that it is a great adsorption applicant.

Thanks to the intersection among conduction and valence band at six positions in momentum at the Dirac points, graphene is often considered a zero-gap semiconductor [6]. Zero band gap depicts "zigzag" with the presence of an "armchair". Descriptive illustration of various structure of graphene is represented in **Figure 1**. Moreover, high charge room temperature durability, as demonstrated by previous research, is 15000 $cm^2.V^{-1}.s^{-1}$ [7]. In some earlier studies, the identical charge flexibility for hole and electron was also stated [8–11]. The splitting stability of charge carriers at a room temperature of audible graphene phonons is noted to be 4.5×10^3 times more than copper [12]. At room temperature, graphene plates give lowest resistivity 10^{-6} Ω cm, which is lower compared to silver [13]. Graphene in twisted bilayer form exhibited superconductivity [8, 14]. Owing to its adsorption power which is nearly 2.3% of the red light, and approximately 2.6% of the green light, the mono-atomic dense bilayer surface can be observed with the naked eye [14]. Graphene has an exceptional clarity for monolayer atomic structure in vacuum [15]. Thermal behavior, one of the key characteristics of graphene, an efficient and desirable tested route for many researchers because of its high ability in thermal applications. In previous experiments, a comparative study of graphite and graphene found that the thermal conductivity varies in both materials at room temperature which are 2000 $W.m^{-1}.K^{-1}$ and 5300 $W.m^{-1}.K^{-1}$, respectively [16]. However, recent progress has demonstrated that the former thermal conductivity value is sustainable, in fact it ranges from 1500 to 2500 $W{\cdot}m^{-1}{\cdot}K^{-1}$ for individual layer of graphene [17–25].

Graphene has two dimensions. The LA and TA showed linear dispersion relation whereas the quadratic scattering relation was observed because of off-plane mode. That is why the linear dispersion mode has high thermal conductivity relative to the off-plane mode [26]. The negative GPs found by the phonon graphene bands and at low temperatures are significant, while the clear link between the negative GPs and thermal expansion coefficient [27]. The graphene structure is layered and the spacing is around 0.335 nm between each sheet. Toughness of graphene is 130 GPa and it has 1TPa Young's Modulus, which is why it is significantly a stronger material than others [28]. Scientists have found that Graphene monolayer have a large-angle-bent, which gives a slight strain, so 2D carbon monoliths displayed important mechanical and physical features. Moreover, the charge mobility in the monolayer of graphene does not alter after high disruption [29].

2.1.1 Applied applications of graphene based materials

Graphene is versatile material owing to excellent physical properties. It has been used in many applications in industry and environment. Following are the applications of components from graphene.

- Energy storage and solar cell [30, 31]
- Sensors, photo detectors, Transistors and memories [32–40]
- Flexible, stretchable and foldable electronics [41–45]
- Biotechnology and medicines [46–51]

2.2 Structural graphite

The word "Graphite" is taken from a Greek-word "graphein" having meanings of "to write". This is a grayish-black naturally-occurring carbon-material with a radiant black-shine. It is a unique material which shows properties of both crystalline & non-crystalline and of a metal & non-metal. On this basis, graphite can be classified as natural and synthetic-graphite.

2.3 Natural-graphite

Naturally-occurring graphite is further grouped into three classes:

- Crystalline-graphite or structural-graphite
- Amorphous-graphite
- Flake-graphite

2.3.1 Crystalline or structural-graphite

Graphite has a layer-structure with hexagonal-arrangement of C-atoms containing covalent-bonding "honeycomb-structure". The layers are stacked together by secondary-bonding type that is, Van-der-Waals interactions which measure the weak shear-strength of graphite. Therefore, by applying a small shear-stress, deformation in structure happens and thus graphite comes out to be anisotropic where properties depend on the direction of applied-force. In structural-graphite, every C-atom is covalently associated to three neighboring C-atoms and that is how each atom leaves a spare free-electron. These free-electrons form a delocalized-cloud of electrons which is weakly bonded to layers which is an ultimate-reason of graphite's good electrical-conductivity along each layer [52]. Structure of graphite is respresented in **Figure 2**.

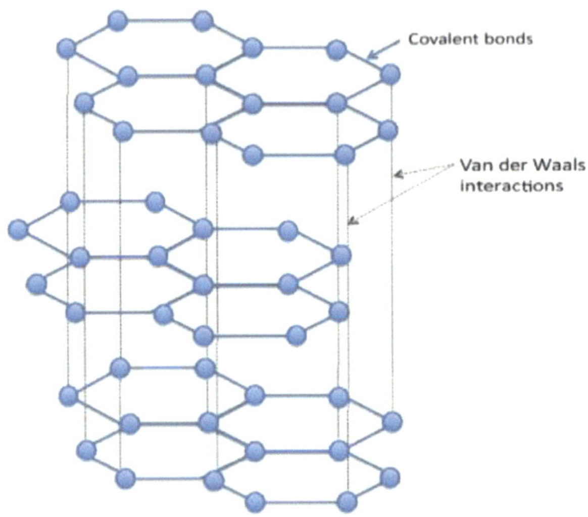

Figure 2.
Structure of graphite [52].

2.3.2 Artificial or synthetic-graphite

Artificial-graphite is obtained by graphitization of nongraphitic carbon and via chemical-vapor-deposition CVD from hydro-carbons using higher temperatures. This graphite is not highly crystalline as natural-graphite or structural-graphite. Example includes synthetic-graphite obtained via heating calcined-petroleum at about 2800°C.

2.3.3 Applied applications and physical properties of graphite

On account of structural, chemical and mechanical properties of graphite, some important applications are listed in the table below:

Properties	**Areas of application**
Chemical-inertness & high-T stability	Used as refractory-materials such as in making of refractory-bricks Mag-Carbon 'Mg-C'.
Directional electrical-conductivity	In chemical-industry, graphite is used as an electrode [53] in specific electrolytic-solutions and in various anti electrical-instrumental coverings as charges-passivator.
High-coefficients of thermal-conductivity & low-absorption of neutron	These properties enable graphite to use as moderator-rods and reflector-components in nuclear-reactors [54].

2.4 Carbon foam (C-foam)

The C-materials are best known for their wide-range porous-structure with variety of size and number of pores [55, 56]. Among fibrous, tubular, granular and other platelet & spherical morphologies of C-materials, C-foam has a particular-pore organization, where there is an interconnection of various macropores (cells) to form an open-cell-structure. Actually, this unique cell-structure defines the novel characteristics and features of these substances such as low density, high thermal-stability, water resistive surfaces, efficient thermal and electrical-conductivities, etc.

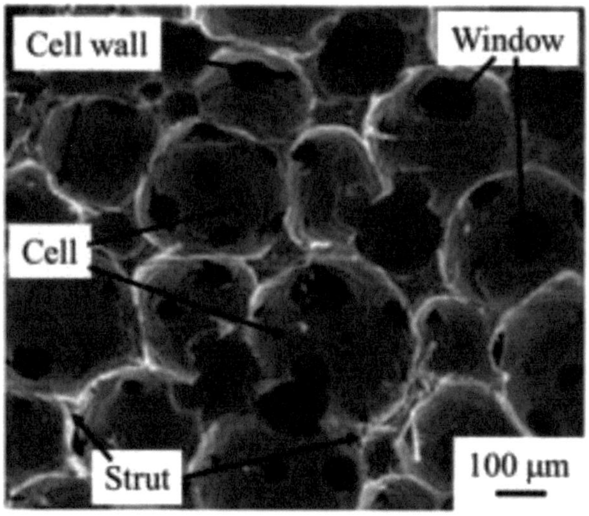

Figure 3.
SEM-shot of C-foam [57].

Thermal and electrical conductivities can be altered for bulky-C-products by making controlled-changes in the internal. Cell-structure to define C-foam can be understood thoroughly from the SEM-image in **Figure 3**. Two distinct pores can be observed in the image: a macropore called "cell" which is surrounded by "C-wall" and other is a hole in C-wall called "window". Window connects the neighboring cells together to form a Cell-structure. With these terminologies, the concept of C-foam is described as: Components with enlarged pores where cells (macropores or sometimes meso-pores) are interconnected through windows and thus provide space to introduce other types of substances inside the pores to increase substrate compatibility [58].

2.4.1 Applied application and physical properties of C-foam

Properties	Applications
Low & High-electrical-conductivity (0.01-10^6 Ω)	Low conductivity-foams are perfect-electrical-conductors having a wide-bandgap to use in radar-absorption and electromagnetic-shielding applications. Similarly, RVC-foams (foams obtained through reticulated-vitreous-carbon) with high conductivity, 97% high-pore volume and honeycomb-structure are used as optically-transparent electrodes [59].
High-thermal conductivity	With good thermal conductivity and low thermal-expansion coefficients, C-foams can be used in thermal-energy storage-devices [60].
Coupling of low-weight, high-strength & resistance to fire	These properties enable C-foam to use in air-crafts and ship materials.
High-specific-area due to open-cell structure	Activation agents are used in C-wall to increase the volume of pores which ultimately provide a larger-specific-area for the adsorption/desorption of water [61–63], radioactive-Cs [61–65], carbon-dioxide and oxygen.

2.5 Carbon nanotubes (CNTs)

As the name suggests, these are tubes of carbon with diameters of nanometer-range which are also known as "bucky-tubes". The tubular C-structure was first studied in 1991 by Iijima [66, 67]. These were named as "multi-walled CNTs

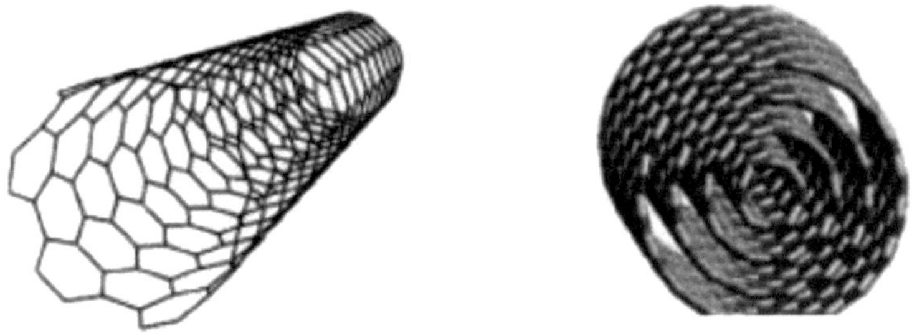

Figure 4.
(a) SW-NT [4] & (b) MW-NTs [71].

(MW-CNT)". Single-partition (SW) CNT was synthesized by Bethune [68]. Both the types can synthesize using three methods: arc-ablation, thermal decomposition and catalytic-growth. To understand the structure of SW-CNT, structure of crystalline-graphite should be known. As graphite contains stacked-layers of hexagonally-arranged C-atoms with sp^2-configuration. The stacking of these layers is due to inter-molecular forces and separately single layer is known as "graphene sheet". Each nanotube comprises of millions of C-atoms and in SW-NT only ten atoms are arranged at circumference and thickness of tube is ~1 atom [69, 70]. The structures of single wall and multi wall nanotubes are shown in **Figure 4**.

2.5.1 Applied applications and physical properties of CNT's

CNTs show a great combination of superlative-mechanical, electrical and thermal properties which is mainly due to sp^2 C-C bonding ability. These properties have opened their ways to the various areas of industrial-applications, some of them are listed below.

Properties	Areas of application
High-stiffness & High-tensile strength	CNTs are the strongest-material among other C-materials ever discovered by humans' history. These are the materials with Young's Modulus >1TPa which is 5× more than of steel and tensile-strength of ~63GPa which is 50× more than that of steel [72]. Due to these properties, CNTs have replaced steel in many applications such as sports-goods: bicycle-components, ice hockey-sticks, baseball-bats, skis etc.
Strong chemical and environmental-reliability & efficient thermal-conductivity	If these properties of CNTs are coupled with the light-weight property of C-based materials, then these materials have potential applications in the field of aerospace [73].
Electrical-conductivity	CNTs have high-electrical conductivity which is comparable to that of copper and are being used in electronics [74].

2.6 Fullerene

The fullerene discovery in 1985 has revotulized the scientific field. It has been used widely in physical, biological and chemical applications, respectively. Amazing physical properties are attribute of C60, which is actually a member of fullerene family.

2.7 Diamond-like carbon (DLC)

Diamond-like carbon (DLC) is undefined carbon [75, 76]. In DLC, considering the hybridization of sp3, no periodicity exists because of the heterogeneity of the bond angle C-C-C. The DLC has low density than that of diamond. Preferably DLC is composed of only sp3 carbon atoms. The processed content, though, sometimes contains sp3 and s2 carbon atoms. The existence of sp2 carbon atoms does not make the substance electrically isolated. In terms of hardness and thermal conduction, DLC is not as strong as diamond, but is much cheaper. The DLC is widely used for painting purposes for better wear resistance and thermal conductivity, owing to its low coefficient of friction. The DLC is normally produced by plasma enhanced chemical vapor deposition (PECVD. The accumulation conditions vary from DLC to diamond. The DLC is hydrogenated frequently during its development to inactive the collar bonds. Such passivation is beneficial to minimize the electric capacitance in the DLC [77].

3. Conclusion

Advanced carbon materials are backbone for scientific revolution of 21st century especially in nanotechnology. Carbon nanotubes (CNTs) and graphene are most researchered nanomaterials for the last decade due to outstanding physical properties. Graphene, being the first ever 2-D material, bring huge opportunities and potential for novel materials research due to good electrical conductivity, thermal conductivity, tensile strength and good dielectric properties, respectively. Due to unique characteristics of graphene monoliths, it has been employed in various fields such as electrical, thermal. Mechanical, and optical applications. In conclusion, advanced carbon materials will be the driving force for next generation scientific revolution.

Author details

Sadia Sharif[1], Sana Arbab[2], Amna Saeed[2], Khurram Shahzad[2], Muhammad Aamir Iqbal[3], Abdullah Khan Durrani[2], Asghari Maqsood[4] and Mujtaba Ikram[2*]

1 Department of Physics, Government College University, Lahore, Punjab, Pakistan

2 Institute of Chemical Engineering and Technology (ICET), University of the Punjab (PU), Lahore, Punjab, Pakistan

3 School of Materials Science and Engineering, Zhejiang University, China

4 Nanoscale Laboratory, Department of Physics, Air University, Islamabad, Pakistan

*Address all correspondence to: mujtaba.icet@pu.edu.pk

IntechOpen

© 2021 The Author(s). Licensee IntechOpen. This chapter is distributed under the terms of the Creative Commons Attribution License (http://creativecommons.org/licenses/by/3.0), which permits unrestricted use, distribution, and reproduction in any medium, provided the original work is properly cited.

References

[1] Boehm, H. P.; Setton, R.; Pure and Applied Chemistry. 66(2012)1893-1906

[2] Simpson, C. D.; Brand, J. Diedrich; Berresheim, Alexander J.; Przybilla, Laurence; Räder, Hans Joachim; Müllen, Klaus 8(2002)1424-1429

[3] Geim, A. Science.324(2009) 1530-1534

[4] Ghoneim, Mohamed T.; Smith, Casey E.; Hussain, Muhammad M. (6 May 2013). AppliedPhysicsLetters.102(2013)

[5] Fraundorf, P.; Wackenhut, M. (2002). Astrophysical Journal Letters.578 (2) 153-156

[6] Jussila, Henri; Yang, He; Granqvist, Niko; Sun, Zhipei Optica.3 (2016) 151-158

[7] Carlsson, J. M. Nature Materials. 6(2007) 801-802

[8] Novoselov, K. S.; Geim, A. K.; Morozov, S. V.; Jiang, D.; Katsnelson, M. I. Grigorieva, I. V.;Dubonos, S. V.; Firsov, A. Nature.438(2005)197-200

[9] Morozov, S.V.; Novoselov, K.; Katsnelson, M.; Schedin, F.; Elias, D.; Jaszczak, J.; Geim, A. Physical Review Letters.100(2008)16602

[10] Chen, J. H.; Jang, Chaun; Xiao, Shudong; Ishigami, Masa; Fuhrer, Michael S. Nature Nanotechnology. 3(2008).206-9

[11] Akturk, A.; Goldsman, N. (2008). Journal of Applied Physics.103 (5) 053702 15. F. V.; Wu, W. M.; Pierpoint, M. P.; Yung, K. C. 67(2014)785

[12] Neto, A Castro; Peres, N. M. R.; Novoselov, K. S.; Geim, A. K.; Geim, A. Rev Mod Phys. 81(2010)109-162 90

[13] Zhu, Shou-En; Yuan, Shengjun; Janssen, G. C. A. M. EPL.108 (2014) 17007

[14] Nair, R. R.; Blake, P.; Grigorenko, A. N.; Novoselov, K. S.; Booth, T. J.; Stauber, T.; Peres, N. M. R.; Geim, A. K. Science.320 (2008)1308-1308

[15] Kuzmenko, A. B.; Van Heumen, E.; Carbone, F.; Van Der Marel, D..Physical Review Letters. 100(2008) 117401

[16] Balandin, A. A.; Ghosh, Suchismita; Bao, Wenzhong; Calizo, Irene; Teweldebrhan, Desalegne; Miao, Feng; Lau, Chun Ning. Nano Letters ASAP.8 (2008) 902-907

[17] Yousefzadi Nobakht, Ali; Shin, Seungha (14 December 2016) Journal of Applied Physics. 120(2016) 225111

[18] Cai, Weiwei; Moore, Arden L.; Zhu, Yanwu; Li, Xuesong; Chen, Shanshan; Shi, Li; Ruoff,Rodney S. Nano Letters. 10(2010)1645-1651

[19] Faugeras, Clement; Faugeras, Blaise; Orlita, Milan; Potemski, M.; Nair, Rahul R. ; Geim, A. K.ACS Nano. 4(2010)1889-1892

[20] Xu, Xiangfan; Pereira, Luiz F. C.et al; Nature Communications. 5 (2014)3689

[21] Lee, Jae-Ung; Yoon, Duhee; Kim, Hakseong; Lee, Sang Wook; Cheong, Hyeonsik Physical Review B. 83(2011)081419. 91

[22] Seol, J. H.; Jo, I.; Moore, A. L.; Lindsay, L.; Aitken, Z. H.; Pettes, M. T.; Li, X.; Yao, Z.; Huang, R.; Broido, D.; Mingo, N.; Ruoff, R. S.; Shi, L. 328(2010)213-216

[23] Klemens, P. G. " International Journal of Thermophysics. 22 ((2001) 265-275

[24] Jang, Wanyoung; Chen, Zhen; Bao, Wenzhong; Lau, Chun Ning; Dames, Chris NanoLetters. 10(2010)3909-3913

[25] Pettes, Michael Thompson; Jo, Insun; Yao, Zhen; Shi, Li Nano Letters. 11 (2011) 1195-1200

[26] Lifshitz, I.M. "Journal of Experimental and Theoretical Physics 22(1952)475

[27] Lee, C.; Wei, X.; Kysar, J. W.; Hone, J.Science. 321(2008)385-388

[28] Briggs, Benjamin D.; Nagabhirava, Bhaskar; Rao, Gayathri; Deer, Robert; Gao, Haiyuan; Xu, Yang; Yu, Bin Applied Physics Letters. 97 (2010) 223102

[29] Zandiatashbar, A.; Lee, G.-H.; An, S. J.; Lee, S.; Mathew, N.; Terrones, M. Hayashi, T.; Picu, R. C.; Hone, J.; Koratkar, N. Nature Communications. 5(2014) 3186

[30] Yin, Zongyou, Jixin Zhu, Qiyuan He, Xiehong Cao, Chaoliang Tan, Hongyu Chen, Qingyu Yan, and Hua Zhang. "Graphene-based materials for solar cell applications." Advanced energy materials 4, no. 1 (2014): 1300574.

[31] Li, Xianglong, and Linjie Zhi. "Graphene hybridization for energy storage applications." Chemical Society Reviews 47, no. 9 (2018): 3189-3216.

[32] Hill, Ernie W., Aravind Vijayaragahvan, and Kostya Novoselov. "Graphene sensors." IEEE Sensors Journal 11, no. 12 (2011): 3161-3170.

[33] Justino, Celine IL, Ana R. Gomes, Ana C. Freitas, Armando C. Duarte, and Teresa AP Rocha-Santos. "Graphene based sensors and biosensors." TrAC Trends in Analytical Chemistry 91 (2017): 53-66.

[34] Wang, Tao, Da Huang, Zhi Yang, Shusheng Xu, Guili He, Xiaolin Li, Nantao Hu, Guilin Yin, Dannong He, and Liying Zhang. "A review on graphene-based gas/vapor sensors with unique properties and potential applications." Nano-Micro Letters 8, no. 2 (2016): 95-119.

[35] Gan, Tian, and Shengshui Hu. "Electrochemical sensors based on graphene materials." Microchimica Acta 175, no. 1-2 (2011): 1.

[36] Schwierz, Frank. "Graphene transistors." Nature nanotechnology 5, no. 7 (2010): 487.

[37] Wang, Han, Thiti Taychatanapat, Allen Hsu, Kenji Watanabe, Takashi Taniguchi, Pablo Jarillo-Herrero, and Tomas Palacios. "BN/graphene/BN transistors for RF applications." IEEE Electron Device Letters 32, no. 9 (2011): 1209-1211.

[38] Schwierz, Frank. "Graphene transistors: status, prospects, and problems." Proceedings of the IEEE 101, no. 7 (2013): 1567-1584.

[39] Wu, Chaoxing, Fushan Li, Yongai Zhang, and Tailiang Guo. "Recoverable electrical transition in a single graphene sheet for application in nonvolatile memories." Applied Physics Letters 100, no. 4 (2012): 042105.

[40] Nashed, R., C. Pan, K. Brenner, and A. Naeemi. "Field emission from graphene sheets and its application in floating gate memories." Semiconductor Science and Technology 33, no. 12 (2018): 125003.

[41] You, Rui, Yu-Qing Liu, Yi-Long Hao, Dong-Dong Han, Yong-Lai Zhang, and Zheng You. "Laser fabrication of graphene-based flexible electronics." Advanced Materials 32, no. 15 (2020): 1901981.

[42] Kim, Sang Jin, Kyoungjun Choi, Bora Lee, Yuna Kim, and Byung Hee Hong. "Materials for flexible, stretchable

electronics: graphene and 2D materials." Annual Review of Materials Research 45 (2015): 63-84.

[43] Sharma, Bhupendra K., and Jong-Hyun Ahn. "Graphene based field effect transistors: Efforts made towards flexible electronics." Solid-State Electronics 89 (2013): 177-188.

[44] Koppens, F. H. L., T. Mueller, Ph Avouris, A. C. Ferrari, M. S. Vitiello, and M. Polini. "Photodetectors based on graphene, other two-dimensional materials and hybrid systems." *Nature nanotechnology* 9, no. 10 (2014): 780-793.

[45] Furchi, Marco, Alexander Urich, Andreas Pospischil, Govinda Lilley, Karl Unterrainer, Hermann Detz, Pavel Klang et al. "Microcavity-integrated graphene photodetector." *Nano letters* 12, no. 6 (2012): 2773-2777.

[46] Zhou, Xianfeng, and Feng Liang. "Application of graphene/graphene oxide in biomedicine and biotechnology." Current medicinal chemistry 21, no. 7 (2014): 855-869.

[47] Hakimi, Mohammad, and Paransa Alimard. "Graphene: synthesis and applications in biotechnology-a review." (2012).

[48] Yang, Mei, Jun Yao, and Yixiang Duan. "Graphene and its derivatives for cell biotechnology." Analyst 138, no. 1 (2013): 72-86.

[49] Feng, Liangzhu, and Zhuang Liu. "Graphene in biomedicine: opportunities and challenges." Nanomedicine 6, no. 2 (2011): 317-324.

[50] Priyadarsini, Subhashree, Swaraj Mohanty, Sumit Mukherjee, Srirupa Basu, and Monalisa Mishra. "Graphene and graphene oxide as nanomaterials for medicine and biology application." Journal of Nanostructure in Chemistry 8, no. 2 (2018): 123-137.

[51] Ding, Xili, Haifeng Liu, and Yubo Fan. "Graphene-based materials in regenerative medicine." Advanced healthcare materials 4, no. 10 (2015): 1451-1468.

[52] Tanzi, M. C., Farè, S., & Candiani, G. (2019). *Organization, Structure, and Properties of Materials.* Foundations of Biomaterials Engineering, *3-103.* doi:10.1016/b978-0-08-101034-1.00001-3.

[53] "Graphite Statistics and Information". USGS. Retrieved 2009-09-09.

[54] Arregui-Mena, J. D.; Bodel, W.; et al. (2016). "Spatial variability in the mechanical properties of Gilsocarbon". Carbon. 110: 497-517. doi:10.1016/j.carbon.2016.09.051.

[55] Zhou P and Chen Q L 2016 J. Anal. Appl. Pyrolysis 122 370

[56] Inagaki M, Konno H and Tanaike O 2010 J. Power Sources 195 7880-7903.

[57] Inagaki M., Qiu J., Guo Q., Carbon foam: preparation and application, Carbon (2015), doi:10.1016/j.carbon.2015.02.021.

[58] Klett J, Hardy R, Romine E, Walls C, Burchell T, High-thermal conductivity, mesophase-pitch-derived carbon foams: effect of precursor on structure and properties. Carbon 2000; 38: 953-973.

[59] Norvell V E, Mamantov G. Optically transparent vitreous carbon electrode. Anal Chem 1977; 49: 1470-1472.

[60] Hoogendoorn C J and Bart G C J 1992 Sol. Energy 48 53-58.

[61] Xiaon N, Ling Z, Wang L, Zhou Y, Qiu J. Preparation of zeolite/carbon foam monolithic composites with a hierarchical structure. New Carbon Mater 2010; 25: 321-324.

[62] Narasimman R, Prabhakaran K. Preparation of carbon foams by thermo-foaming of activated carbon powder dispersions in an aqueous sucrose resin. Carbon 2012; 50: 5583-5593.

[63] Tanzi, M. C., Farè, S., & Candiani, G. (2019). *Organization, Structure, and Properties of Materials.* Foundations of Biomaterials Engineering, *3-103.* doi:10.1016/b978-0-08-101034-1.00001-3.

[64] "Graphite Statistics and Information". USGS. Retrieved 2009-09-09.

[65] Arregui-Mena, J. D.; Bodel, W.; et al. (2016). "Spatial variability in the mechanical properties of Gilsocarbon". Carbon. 110: 497-517. doi:10.1016/j.carbon.2016.09.051.

[66] H.W. Kroto, J.R. Heath, S.C. O'Brien, R.F. Curl, R.E. Smalley, C60: buckminsterfullerene, Nature 318 (1985) 162.

[67] S. Iijima, Helical microtubules of graphitic carbon, Nature (London) 354 (1991) 56.

[68] S. Iijima, T. Ichihashi, Single-shell carbon nanotubes of 1-nm diameter, Nature (London) 363 (1993) 603.

[69] Saether E, Frankland SJV, Pipes RB. Transverse mechanical properties of single walled carbon nanotube crystals. Part I: determination of elastic moduli. Compos Sci Technol, 63, 1543 (2003). http://dx.doi.org/10.1016/S0266-3538(03)00056-3.

[70] Dresselhaus MS, Lin YM, Rabin O, Jorio A, Souza AG, Pimenta MA, Saito R, Samsonidze G, Dresselhaus G. Nanowires and nanotubes. Mater Sci Engg: C, 23, 129 (2003).

[71] T. Lin, V. Bajpai, T. Ji, and L. Dai, "Chemistry of carbon nanotubes," Australian Journal of Chemistry, vol. 56, no. 7, pp. 635-651, 2003.

[72] R. Saito, G. Dresselhaus, and M. S. Dresselhaus, *Physical Properties of Carbon Nanotubes*, London and Imperial College Press, 1998.

[73] M. F. Yu, B. S. Files, S. Arepalli, and R. S. Ruoff, "Tensile loading of ropes of single wall carbon nanotubes and their mechanical properties," Physical Review Letters, vol. 84, no. 24, pp. 5552-5555, 2000.

[74] J. A. Elliott, J. K. W. Sandler, A. H. Windle, R. J. Young, and M. S. P. Shaffer, "Collapse of Single-Walled Carbon Nanotubes is Diameter Dependent," Physical Review Letters, vol. 92, no. 9, pp. 1-4, 2004.

[75] Robertson, John. "Classification of diamond-like carbons." In *Tribology of Diamond-Like Carbon Films*, pp. 13-24. Springer, Boston, MA, 2008.

[76] Ohgoe, Yasuharu, Kenji K. Hirakuri, Hidetoshi Saitoh, Takahiro Nakahigashi, Naoto Ohtake, Atsushi Hirata, Kazuhiro Kanda, Masanori Hiratsuka, and Yasuhiro Fukui. "Classification of DLC films in terms of biological response." Surface and Coatings Technology 207 (2012): 350-354.

[77] Grill, A. "Diamond-like carbon coatings as biocompatible materials—an overview." Diamond and related materials 12, no. 2 (2003): 166-170.

Chapter 2

Advanced Carbon Materials: Base of 21st Century Scientific Innovations in Chemical, Polymer, Sensing and Energy Engineering

Muhammad Ikram, Ali Raza, Khurram Shahzad, Ali Haider, Junaid Haider, Abdullah Khan Durrani, Asim Hassan Rizvi, Asghari Maqsood and Mujtaba Ikram

Abstract

Advance carbon material that includes graphene, fullerenes, hierarchical carbon, and CNTs are referred to as strength of revolution and advancement in the era of material science and technology. In general, 20th century corresponds to plastic meanwhile 21st century will be named as "Century of Graphene" owing to its exceptional physical properties. Graphene is now well-known and prominent 2D carbon allotrope that is considered as multipurpose material in comparison with any material discovered on earth. One of the interesting properties of graphene is strongest and lightest material that enables it to conduct electricity and heat as compared to any other material. Such features permit it to utilize in numerous applications including biosensors, electronic industry, environmental remediation, drug delivery, energy storage, and production as well. Owing to these capabilities, it can be stated that graphene can be utilized to improve effectiveness and performance of existing substances and materials. In the future, conjugation of graphene with other 2D material will be devolved to produce further remarkable compounds that make it appropriate for an extensive variety of applications. This chapter grants the utilization and applications of advanced carbons materials in chemical, polymer, sensing and energy enegineering.

Keywords: polymer composites, nano coatings, lubricants, nanofluids, biosensors, fuel cells, supercapacitors

1. Introduction

Carbon has been distinguished into variety of forms as amorphous carbon, diamond, and graphite. Among these, the well-recognized allotropes of carbon since ancient times are diamond and graphite. The third kind of carbon named fullerene was discovered by Kroto et al. in 1985 whereas; carbon nanotubes (CNTs) were discovered by Iijima in 1991 that leads to gain a significant role in the field of science and technology. Accordingly, only three kinds of carbon allotropes were identified and well-known in the carbon family are first, diamond and graphite

(3D) secondly, CNTs (1D) and thirdly, fullerenes (0D). Later, in 1991 it was realized that CNTs were fabricated by rolling of 2D graphene single sheet that was extracted from 3D graphitic material [1–3]. Furthermore, isolation of graphene was somewhat struggling and indefinable concerning any effort corresponds to experimental research until 2004. Graphene is an elementary structural element of CNTs, fullerenes, and graphite that are named as allotropes of carbon family. Fullerene is entitled as buckyball as it is composed of carbon sheets in the arrangement of spherical profile. In comparison with fullerene, CNTs acquire tubular form. For more than two decades, CNTs and fullerene-based martial expose extensive applications in the varied portion of the research that involve biosensors, super-capacitors, electrochemical sensors, electronics, fuel cells, batteries, and medicinal applications. Presently, graphene is entitled as "Rising Star Candidate" after its effective production from scotch tape process by utilizing voluntarily accessible graphite by Andre Geim and his coworkers in 2004. Single-layer sheets of graphene consist of carbon atom that is sp^2 bonded and acquires honeycomb-like lattice which is densely packed. As an active material, remarkable properties of graphene that include tunable bandgap, high specific surface area, superior thermal, electrical stability and conductivity, and more importantly Hall effect (at room temperature) provides suitable platform for its utilization in the production of several composites materials [4]. Struggles were devoted to reviewing the structure and preparation of graphene its properties, possible applications, and finally composite material [5–8]. At present, owing to possess remarkable properties, graphene is shortlisted as the most widespread material that can be employed for several devices and applications. This chapter grants the utilization and applications of graphene in various approaches, the synthesis routes, and numerous exceptional properties.

2. Application

Previously, graphene has illustrated promising impression to various information communication technology areas that sorts from a high-performance application (top-end) in ultrafast information processing (i.e. THz) to consumer applications by means of flexible electronic structures. An authentic property of graphene is verified by the increment in the score of chip makers now energetic in research based on graphene. Prominently, graphene is reflected as the emerging candidate that can be utilized for post-Si-electronics. Most auspicious applications of graphene contain light processing, sensors, electronics, plasmonics, energy storage, meta-materials, generators, etc. Besides, graphene is utilized to enhance various industrial and medical processes. The overview for the applications of graphene is displayed in **Figure 1**.

2.1 Polymer composites

Biphasic materials are considered as polymeric composites, which are attained by dispersing one phase into another controllably. Modified graphene may be dispersed into polymer-matrix to become reinforcing-filler to increase optimally physiochemical properties [10]. Firstly, Stankovich et al. presented phenyl isocyanated graphene acting as nanofiller during the synthesis of polystyrene (PS)/graphene matrix [11]. It was observed that only 2.4 vol% increments belonging to surface-modified graphitic compound filled desired composites, caused by enlarged surface area graphitic composite. The electrical conductivity attained percolation threshold by incorporating about ~0.1 vol% graphene as illustrated **Figure 2**. Reports offered by Eda et al.

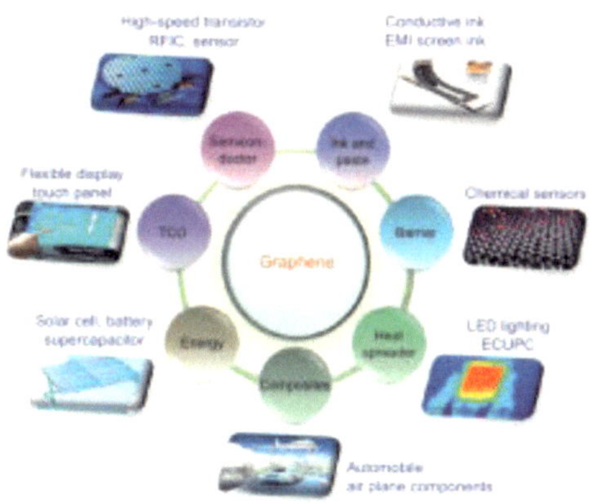

Figure 1.
Overview of applications of advanced carbon material (graphene) [9].

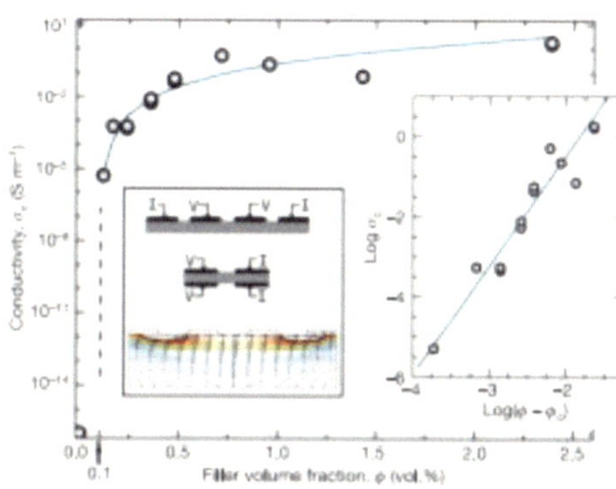

Figure 2.
Electrical conductivity of phenyl isocyanate modifi ed. graphene filled polystyrene composites [11].

exhibited functionalized graphitic filled PS-composites showing the same electrical properties as that of monolayer rGO nanosheets [12]. Whereas PS-composites are being explored p-type semiconducting nature at high temperatures.

Kuila et al. reported that dodecyl-amine (DA) along with octadecy-amine (ODA) functionalized graphitic filler during synthesis of linear-low-density polyethylene (LDPE), while ethylene-vinyl-acetate (EVA) composites were obtained respectively [13–16]. Modified-graphene was well dispersed in LLDPE as well as EVA matrix during hydrophobic-interaction with alkyl-chains of polymer matrix along with modifier. As far as tensile strength is concerned, it acts with storage-modulus collectively to composites as they are increased with surface-modified graphene optimally to a certain limit, thereby decreasing with additional fillers. Epoxy graphitic composites have been detailed in investigation processes [17–19]. It has been keenly observed that little increment of surface-modified graphitic material increasingly enhances mechanical as well as thermal stability as compared with neat epoxy. This corresponds to reasonable surface area along with superior mechanical-strength attributing to graphene composite. Some other usable polymers for graphitic-composite preparation are known as polyvinyl-alcohol, crystal-polymers, polypropylene, polypyrrole, polymethyl methacrylate chitosan, cellulose, polycarbonate, polyethylene terephthalate, and polyvinyl chloride [20–25]. Graphitic-polymers offer potential applications towards automobiles, air-craft industry, turbine blades, bony structures, and tissue culture implantations [26–30].

2.2 Nano coatings: antimicrobials and microelectronics

Carbon nanotubes show a promising multifunctional nature material in coating fabrication. Metal decorated CNTs are considered as hybrid-systems that may be prepared by using those CNTs having carboxyl-groups, binding transition metal ions such as Ag^{+} and Cu^{2+}. These aforesaid ions contribute a large part to their superior antimicrobial activity to destroy bacterial as well as fungus microbes along with less cross-resistance towards antibiotics (see **Figure 3a** and **b**) [31]. MWCNTs with paint materials successfully reduce biofouling-ship-hulls owing to discourage embodied algae with barnacles [32]. Consequently, they are referred to as alternatives for environmentally polluted biocide type paints. Anticorrosion-coatings include CNTs for metals for enhancement of coating-stiffness and strengthen them to make an electric pathway to create cathodic protection.

Widespread progress has been made for fabrication of CNTs that are based on flexible and transparent conductive thin films [33–35] proving alternative material of indium tin oxide. The main issue concerning ITO is its expensive nature owing to shortage of indium. However immense need for displays, touchscreens and photovoltaic provide stimulus. Moreover, CNTs flexibility raise the transparency of conductors showing a major advantage over ITO coatings towards flexible displays. Additionally, transparent CNTs conductors are deposited from solutions such as slot-die coating as well as ultrasonic spraying along with cost-effective non-lithographic approaches likewise micro-plotting. The latest effort has been made for fabrication of CNTs films showing 90% transparency with 100-ohm resistivity per square as is clear from **Figure 4**. Surface resistivity so much appeared is considerably suitable for promising applications. However, substantially it is better than equally transparent and optimal doping with ITO coatings [36]. Widespread applications have exhibiting requirements relevant to CNTs thin-film-heaters and are substantially used for defrosting automobile-windows as well as sidewalks. Aforesaid all types of coatings are widely used on an industrial level.

Recently, CNTs films are transparent; however, stretchable flexible may often tailor in the form of shapes and sizes. They are freestanding and are placed on rigid or flexible insulated surfaces. A piece of carbon nanotube CNTs thin films may explore magnet-free-loudspeaker. It may simply show through applying an audio-frequency-current passing through it depicted in **Figure 5**. CNTs film loudspeaker produces sound waves with high-frequency range, a wide range of sound pressure-level along with low harmonic-distortion [37]. These CNTs thin films behave like transistors, proving more attractive towards driving organic light-emitting-diode

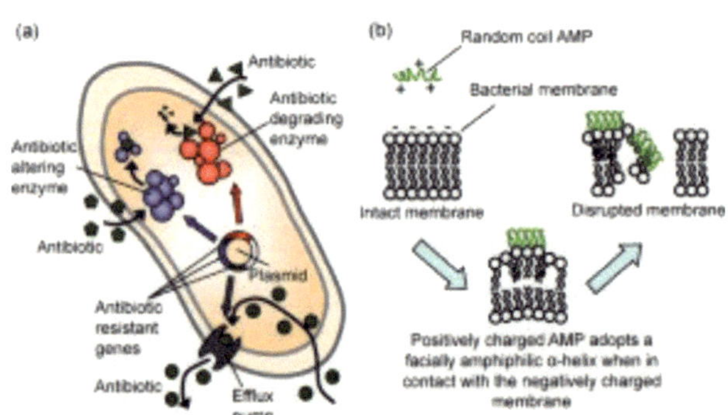

Figure 3.
*Comparison in functional mechanism between small molecular antibiotics and macromolecular antimicrobials (**a**) mechanisms of antibiotic resistance in bacteria, (**b**) mechanism of membrane-active antimicrobial peptides.*

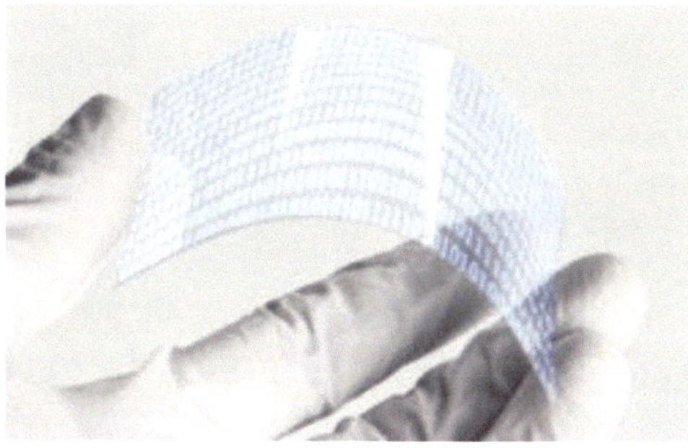

Figure 4.
Carbon nanotubes flexible transparent conducting film ((image courtesy Plasticstar material news).

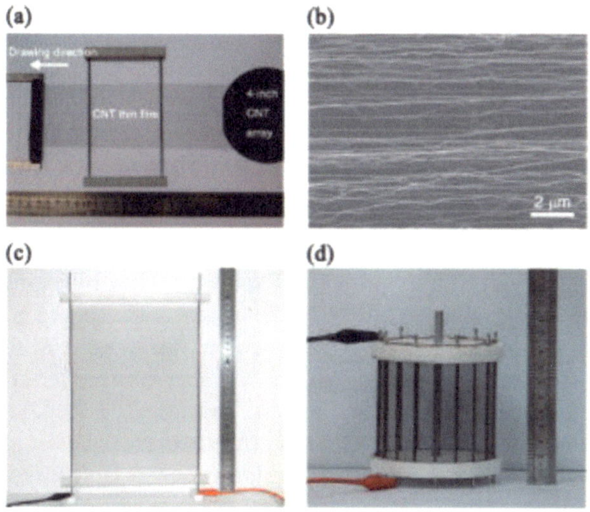

Figure 5.
*Carbon nanotube thin film loudspeakers **(a)** the CNT thin film was pulled out from a super aligned CNT array grown on a 4 in. Silicon wafer and put on two electrodes of a frame to make a loudspeaker. **(b)** SEM image of the CNT thin film showing that the CNTs are aligned in the drawing direction. **(c)** A4 paper size CNT thin film loudspeaker. **(d)** the cylindrical cage shape CNT thin film loudspeaker can emit sounds to all directions, diameter 9 cm, height 8.5 cm [39].*

screen display. Because of this reason, they have explored higher-mobility as compared with amorphous silicon, depositable by low-temperature, and vacuum-free approaches. Today flexible CNTs -TFTs having mobility 35 $cm^2 V^{-1} s^{-1}$, whereas an on/off ratio of 6×10^6 has been demonstrated in **Figure 5a, d** [38].

2.3 Lubricants

Applications corresponding to surface-functionalized-graphene show additive counterpart in lubricant oil refinery owing to progressing research field. Extremely large mechanical-flexibility, fine friction-reduction, greater surface-area, and anti-wear-ability support enhancement in properties. In addition, Zhang et al. also observed oleic acid-modified graphitic nature lubricant [40]. Tribological properties were investigated by employing four-ball tribometer relevant to oily surface-modified graphene. **Figure 6a, b** illustrates lubricant optimized-graphene with contents (0.02–0.06 wt.%), exhibiting improved-friction as well as anti-wear activity, 17% friction-coefficient whereas 14% wear scar-diameter respectively. Desired friction behavior has been elaborated by

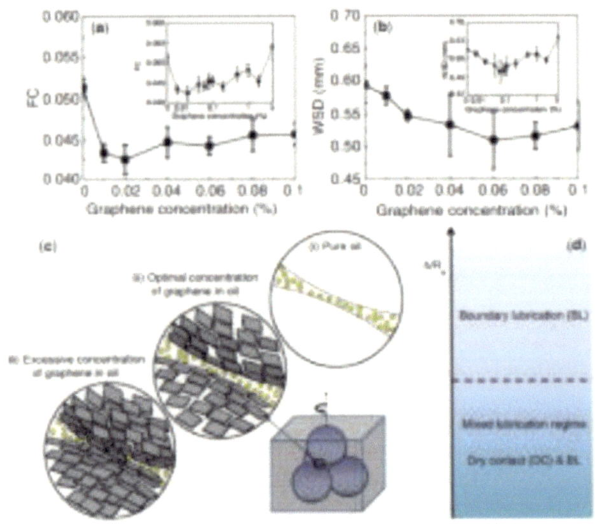

Figure 6.
Four-ball test results: ***(a)*** *FC versus graphene concentration;* ***(b)*** *WSD versus graphene concentration;* ***(c)*** *schematic diagram of the tribological mechanism of graphene sheets as oil additives;* ***(d)*** *lubrication regime transition [40].*

proposed tribological activity as shown in **Figure 6c**. Graphitic protective-layer became prominent on each steel ball surface separately with less concentration, thereby introducing improved anti-wear performance. On the other hand, oily films become discontinuous with higher density that is considered responsible for antiwear-degradation properties. Lin et al. investigated (0.075 wt%) stearic with oleic-acid modified graphitic nature in oil tunes wear-resistance along with load-carrying machine efficiency [41]. Current reports also presented that alkylated graphitic organic solvents may show lubricant behavior to improve properties [42]. Alkylated-graphene with different alkyl-chain-length (Cn = 8, 12, 18) is synthesized by condensed medium reaction (alkylamine+$SOCl_2$-activated GO). It was investigated through octadecyl amino-graphene mixed with hexadecane. In this case, reduced friction along with wear concentration (26% and 9%) was obtained compared with hexadecane.

2.4 Nanofluids

Loss of energy in the form of heat energy slows down performance of various instruments and mechanical technology. Instrument and machinery performance may be improved by using some fluids such as DI water, transformer oil, and heat-sensitive fluids. Heat transfer capability of fluids is less enough caused by the deterioration of productivity and lifetime of equipment and machines and also electronic circuits. To prolong heat transfer efficiency, the addition of nanomaterials addition is increased in fluids that may further improve the efficiency. Baby et al. reported thermal conductivity that may be increased upto14% with temperature (25°C) and deionized water is used as base-fluid showing fraction by volume of only 0.056% [43, 44]. Moreover, thermal conductivity is increased upto 64% at 50°C but with same contents belonging to modified-graphene. Ghozatloo et al. observed 0.06 wt% functionalized-graphene may improve thermal-conductivity (14.2%) when treated in water (25°C) [45]. Finally, thermal-conductivity is enhanced (18%) when the temperature is increased to 52°C.

2.5 Graphene-based transparent and flexible conductive films for displays and electrodes

Graphene is incorporated into electronics field by employing transfer printing along with solution-based approaches. Chhowalla et al. [46] suggested an efficient approach for smooth deposition with effective control of reduced graphene oxide in the form of thin films having thickness like single-monolayer to several-layers ranging large areas. Optoelectronic properties are tuned over reasonable order of magnitude that presents potentially beneficial towards transparent semiconductors as well as semi-metals. The thinnest films show graphitic ambipolar-transistor behavior. However, thick films behave like graphitic semi-metals respectively [47, 48]. Consequently, suggested deposition in this approach offered new routes to translate fundamental properties relevant to graphene into viable devices. Furthermore, large-scale transparent electrode growth has been successfully presented by Hong et al. [49] In this work, chemical vapor deposition technique was employed on thin nickel films. Two methods were applied for the formation of films and thereby transferring also to arbitrary substrates. Graphene films exhibited sheet resistance as well as optical transparency at desired level respectively. Graphene monolayers were transferred to SiO_2 substrates showing electron mobility at faster rate along with half-integer (quantum Hall effect). High-quality graphene was grown by CVD that presented better results as compared with mechanically cleaved graphene as illustrated in **Figure 7a-c**. Owing to extraordinary mechanical properties, graphene demonstrated macroscopic use upto maximum level, resulting in conducting electrodes and transparent electrodes in (flexible and foldable) electronics [50].

In addition, superior optical and electronic graphene properties i.e., high mobility, optical behavior, flexibility trend, and environmental stability are accounted for promising material attributing to applications towards photonic as well as optoelectronic fields. In this support, comprehensive literary work has been done favorable for graphene photonics, optoelectronics, and other applications were offered by Ferrari et al. [51]. From scientific contents included in the review clearly show graphene-based conducting films and graphene oxide (GO) based conducting films that were used in synthesis of various photonic with optoelectronic devices. Equipment such as inorganic and organic electrodes of dye-sensitized solar cells, light-emitting diodes as well as electrochemical cells, touch screens, graphene-based absorbers.

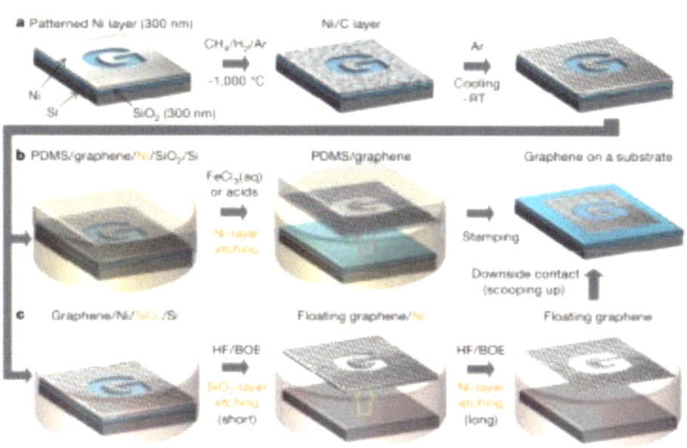

Figure 7.
*Synthesis, etching, and transfer processes for the large scale and patterned graphene films, **(a)** synthesis of patterned graphene films on thin nickel layers **(b)** etching using $FeCl_3$ (or acids) and transfer of graphene films using a PDMS stamp **(c)** etching using BOE or hydrogen fluoride (HF) solution and transfer of graphene films.*

Graphene electrodes showing high-performance field-effect transistors have been fabricated by Kim et al. [52]. To optimize performance of devices, authors controlled work-function attached with graphene electrodes via functionalization of SiO_2 substrate surface. NH_2 may donate electrons that are considered terminated SAMs but they are induced n-doping strongly in graphene. On the other hand, CH_3-terminated SAMs contributed neutralized p-doping that was strictly induced through SiO_2-substrates. Resultantly, graphene electrode work function considerably changed. Moreover, SAMs were observed as pattern-able robust yield. Besides, output of work may also be used towards fabrication of various graphitic nature compounds that paved foundation of electronic as well as optoelectronic devices.

Graphene films indicate mechanical along with optical properties as compared with other transparent-thin-films, particularly in photonics and optoelectronics. However, as far as conductivity is concerned it is inferior as compared with conventional (ITO) electrodes having comparable transparency and resulting in lower performance of devices working on graphene-based transparent thin films. Ahn et al. [53] presented an effective method to overcome deficiency and to improve graphene films concerning performance towards electrostatically doping that was employed through ferroelectric polymer. Aforesaid graphene films showing ferroelectric polarization have been used for the preparation of ultrathin organic-solar-cells (OSCs). Graphene-based OSCs have explored superior efficiency as well as superior stability as compared with graphene-based OSCs that were chemically doped. Moreover, OSCs fabricated by ultrathin-ferroelectric-film act as substrate with few micrometer sizes, exhibited attractive mechanical flexibility as well as durability. In the last, these may also be rolled up into cylindrical shapes having 7.5 mm diameter size.

2.6 Graphene-based separation membranes

Graphene nanopores sheets are used as separation membranes emerging and covering various since theoretical studies that were presented by Král et al. [54]. They were labeled modified nanopores incorporated graphitic type monolayers thereby resulted from molecular dynamic-simulation providing superior realm of hydrated ions. The ions in a partly stripped state connected with hydration shells may penetrate through infinitesimal pores having diameter ~5 Å, such as fluorine with nitrogen terminated-pores permit flow of Li^+, Na^+ and K^+ like positive ions having ratio 9:14:33 systematically whereas negative ions are strictly prohibited. On the other hand, hydrogen-terminated pores accelerate F^-, $Cl^-$and Br^- anions along with a specific ratio 0:17:33 rather it blocks cationic passage. Aforesaid nanopores may provide versatile promising applications, particularly towards molecular separation and energy storage devices respectively.

In addition, Jiang et al. [55] contributed the work that dealt with permeability as well as selectivity related to graphene sheets structured with nanometer-scale pores adopting density functional theory for necessary calculations. Researchers investigated superior selectivity order of magnitude that was 10^5 for H_2/CH_4 showing excellent performance from H_2 side in the situation of nitrogen-treated pore. Furthermore, report writers investigated selectivity at an extremely higher order of magnitude equivalent to 10^{23} for H_2/CH_4 for all hydrogen functionalized pores with width infinitesimally 2.5 Å, presenting a barrier (1.6 eV) for methane (CH_4) whereas surmountable for H_2 with magnitude 0.22 eV. These results exhibited that pores are considered superior to polymers as well as silica membranes. Whereas bulk solubility along with diffusivity is plays a dominant role to transport gas molecules throughout the material. Outcomes suggested one atom thin porous-graphene-sheets behave such as highly efficient and selective membranes relevant to gas separation. Aforesaid types of pores may occupy a widespread impact concerning various energy devices with technological applications.

The molecular-dynamic-simulation employed by Xue et al. [56] explored CO_2 separation strategy from that of CO_2 mixture whereas N_2 gas through porous graphene-membranes. Graphene sheets are chemically functionalized to observe its effects while porous graphene membranes performance for separation has been controllably examined. Researchers investigated chemical functionalization of graphene sheets that may increase absorptive capability of CO_2 gas. On the other hand pore-rim chemical-functionalization significantly enhanced CO_2 selectivity over N_2 gas molecules. The results demonstrated versatile use of functionalized-porous-graphene for CO_2 as well as N_2 separation. Resultantly authors suggested an effective strategy, improving gas separation activity of porous-graphene-membranes [57].

Nanoporous graphene use for water desalination has been suggested by Grossman et al. [58]. Through employing classical-molecular-dynamics, this work presented nanometer-scale porous single-layer-graphene that may prove filter of (NaCl) effectively from that of water. Furthermore, authors researched desalination-performance corresponding to membrane exploring functioning of pore-size, chemical-functionalization as well as applied-pressure. The results indicate membrane's ability that prevents salt penetration and all depends on the porous-diameter size along with sized pores suitable for water flow whereas passage of ions was banned. Further investigation indicates role of functional-groups appeared at graphene-edges in hydroxyl group may form commonly double hydrophilic nature. However, water flux is increased taking place by the reason of salt rejection activity with less amount corresponding to capability of hydroxyl group substituting water molecules in hydration-shell of ions. Collective and achieved outcomes that explored water-permeability of relevant material were clearly in higher magnitude as compared with reverse-osmosis membranes conventionally, thereby NPG may perform valuable role play towards water refinement [59, 60].

The same period was covered by Karnik et al. [61] study also who selectively suggested transport of molecules employing intrinsic-defects single-layer (CVD) graphene. In this case, small measured area was identified greater than 25 mm^2, but in turn it was transferred on porous polycarbonate-substrate. The collective contribution of pressure-driven as well as diffusive- transport with precise-measurement presented confirm evidence with respect to size-selective- transport of material molecules passing through membranes. They were attributed to low-frequency presence within 14 nm range diameter size pores relevant to (CVD) graphene as describe in **Figure 8**. Consequently, authors have proposed first step towards the occurrence of graphene-based selective- membranes [62–65].

Previous work was progressively continued [66] for molecular-sieving by employing porous- graphene. In this respect, Bunch et al. [67] also fabricated valves to control gas-phase-transport through graphene containing discrete nano-sized pores. Reports have revealed and identified gas-flux passing through discrete nano-size pores present in monolayer-graphene that may be detected as well as controlled employing nanometer-size gold clusters. These clusters are centered on graphene surface by migrating pores but partially block them also. However, samples containing not gold-clusters indicate stochastic-switching of magnitude of gas molecules attributing rearrangement of desired pores. Additionally, previously fabricated molecular valves may be involved particularly to progress ideal approaches towards a molecular synthesis that are considered foundation for controllable switching concerned with molecular gas flux [68, 69].

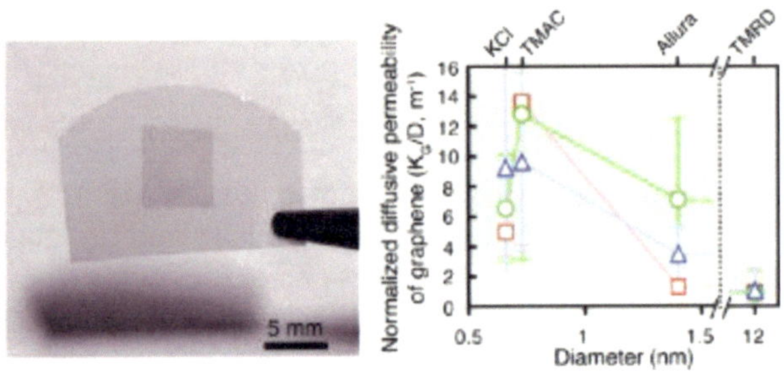

Figure 8.
(a) *Graphene composite membrane (GCM) consists of large-area graphene on polycarbonate track etch (PCTE) membrane,* ***(b)*** *permeability of the CVD graphene, KG, calculated for the three membranes using a simple circuit model (inset), indicated as a function of the diameters of the molecules. Only two pores, one of which is covered by graphene, are shown for clarity. The gray region denotes the continuum model prediction for graphene of porosity between 0.025% and 0.15% [61].*

2.7 Biosensors

Sensors are regarded as those devices that may identify changes in occurring events. Various studies have reported CNTs to use concerning sensors such as chemical, thermal, biological, and gas respectively. In addition, CNTs may also behave like flow sensors [70, 71]. It has been observed that liquid flow on SWCNTs bundles creates voltage normally in flow direction, and may be used in near future in the form of micro-machines working in a fluid medium, for example, heart pace-makers working without heavy-battery as well as recharging [70]. Piezoresistive sensors based on pressure may be prepared using CNTs. SWCNTs have also grown on polysilicon membranes [72]. Uniform pressure creates change into resistance of SWCNTs that was observed in membranes. From viewpoint of Caldwell et al. [73] piezoresistive fabrication offered pressure sensors for CNTs that may bring changes dramatically to biomedical industry and various piezoresistance diagnostic nature as well as therapeutic devices have recently applied in sensor field. Moreover, CNTs fabricated biosensors are used to detect deoxyribonucleic acid concentration in the body. Aforesaid instruments also detected specific parts of DNA corresponding to particular type of disease [74]. Sensors previously mentioned become capable to detect only few molecules of DNA containing specific sequences, thereby increasing probability to diagnose patients possessing specific sequences that are closely related to cancerous genes. Furthermore biosensors have been suitably used for the sensing of glucose. CNTs chemical-sensors, especially for liquids, may also use sensing capability to investigate blood completely or partially. In this case, biosensors are proposed favorable to detect sodium as well as to find pH value accordingly [75].

Having small size with owing attractive electrochemical properties, carbon nanotubes contribute a great part as a component of biosensors. Additionally, CNTs fabricated electrodes possess interesting electrochemical properties as compared with previously available electrodes and show superior quality [76]. CNT-based biosensors present a high aspect-ratio that enables tubes to become embodied into proteins so that electron transferring included with enzymes frequently occur such as glucose oxidase where redox centers are observed not normal to be accessible (See **Figure 9**) [78]. Moreover, chemically modified CNTs have become an effective approach to contribute selectivity property into resulting biosensors that have sufficiently exploited towards exploring sensitivity to detect DNA molecules [79]. However, in near future, fine efforts may be expected to direct towards preventing biomolecules that may be absorbed on surface of tube walls, whereas promising advances have previously contributed a great part in this respect [80]. Further

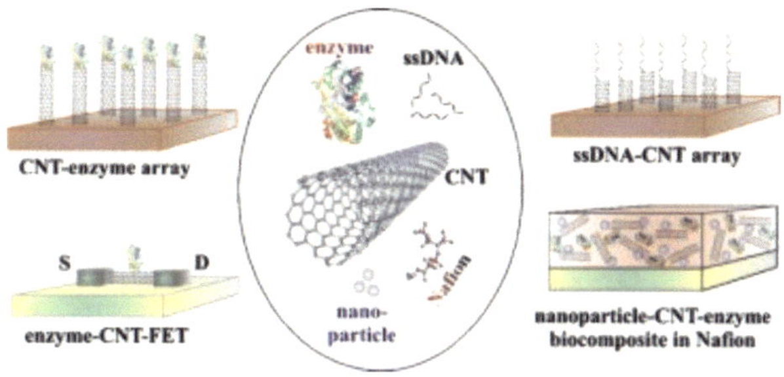

Figure 9.
Schematics of synthesis process DNA based biosensors [77].

advancements may extend range of molecules to be modified that are considered attachable to nanotubes whereas enzymes, as well as nucleic acids along with some metal nanocrystals, are numerously employed to meet the need so far. Particularly respect is electropolymerized coatings have been appreciated that may be prepared with various concentrations, having precise and controllable thicknesses [81].

2.8 Fuel cells

As far as fuel cells are concerned, they are utilized for conversion of chemical energy into electricity directly with great efficiency and exhibited excellent results towards different applications [82–84]. In the fuel cells, catalysts on membrane surface are especially PEM made from graphene. Recently numerous investigations are under progress assessing probability for substitution of platinum catalyst with metals or metal oxides and with nitrogen functionalized metal catalyst [85, 86]. However, some catalysts face issues such as stability as well as activity as compared with platinum catalyst. Active carbon exhibits capability for meeting said challenges yet they tend to occupy certain limitations accordingly. They possess high surface area owing to have instability thereby raising major issues unless coupled with suitable material for this purpose. Graphene technological development made by active carbons has suggested stronger substitutes to platinum occupying high conductivity whereas surface area is considered high along with adhesion property for the catalyst [87, 88]. Graphene oxide, a derivative of graphene resides large number of functional groups making them best for nucleation sites such as catalyst nanoparticles randomly locate on and edges of the surface [89]. The extensive use of graphene is indicated in fuel cells showing supporting material to anode catalyst and replace also cathode catalyst as well as standalone electrolyte membrane and bipolar plates. All work may be summarized concerning role of graphene in various component forms. Platinum, as well as alloys, are supposed as conventional catalysts in the electrodes of fuel cells. They are either an anode or cathode located in fuel cells. These fuel cells are fed by hydrogen and other hydrocarbon methanol [90] as well as ethanol [91]. Platinum is expensive as well as limited in availability and also caused by the produced intermediates while oxidation reactions are carried out at different fuels [92]. Various approaches were employed to reduce catalyst loading or complete replacement of Pt catalyst by using non-precious catalyst reactions at anodes [93] as well as cathode [94] terminals of fuel cells.

2.9 Supercapacitors

Supercapacitors harvest excellent properties such as energy density, ultra thinness, and long life, and therefore have proven promising candidates in electrochemical energy-storage systems [95–97]. Initially, supercapacitors may be categorized into electrical double-layer as well as Pseudo-capacitors depending on energy-storage mechanisms. In first category, charges accumulate electrostatically at electrode and electrolyte interface through formation of an electrical double-charges layer. Charge-storage is uniquely physical essence showing no chemical reaction yet is called non-faradaic process. Electric-double-layer behaves like dielectric whereas capacitance proves direct-function owing to surface-area of electrode. Therefore, carbon-based nanomaterials possessing great surface-area for electrodes increase capacitance of electrical double-layer capacitors. Charge –discharge functioning is indicated by ion absorption-desorption capability of EDLC. Ions are directed forming EDL at the time of voltage application at electrodes which in turn charge EDLC for controlling purpose. It has been observed that carbonaceous electrodes exhibit fine electrochemical surface-area inheriting large porosity caused by creating enhanced interfacial-area forming prominent EDL. Carbonaceous materials have attractive electrical properties owing to which are labeled as basic type of EDLC [98]. Unlike EDLC nature, Pseudo capacitors show capability of fast (Faradaic) charging with transfer-reactions that are carried out at solid electrodes as well as electrolytes. As a result, faradaic-charge-transfer is an applied voltage-dependent system. Fundamental electrochemical reactions in pseudocapacitance involve chemisorption along with electro-sorption from electrolyte. Redox (oxidation and reduction) reactions attractions from electrolyte thereby producing intercalation/de-intercalation sites relevant to active electrodes. Previous electrochemical processes are proposed as surface dependent. In order to promote electrochemical properties attributing to capacitors, great efforts were devoted to making functionalization/hybridization related to a variety of materials or nanostructured optimized promising candidates [99].

3. Conclusions and future directions

Advanced carbons materials such as graphene and CNTs are considered key merits for affordable energy conversions and storage versatile applications. The investigation explored the latest technological advancement during synthesis of the said advanced materials whereas characterizations are performed with respect to current day applications. CVD technique often leads to production of nanostructures having porous networks showing good conductivity. Since quality improvement is the main goal of research work of relevant material, so is improved significantly through employing such technique. Growing concerns are also expected concerning scalability adopting this approach reasonably. Furthermore, characteristics and performance are achievable towards graphene as well as graphene oxide equally growing concern size with quality of graphene-oxide-precursor. The investigation related to novel techniques are aimed to enhance into inter-sheet- binding is considered another novel direction towards research purpose. Desired characteristics are proposed to be achieved through merging graphene sponges as well as polymers. As far as research-based graphene applications are concerned, they belong to several energy storage/conversion devices that are considered still novel in research activities. Graphene suitability has exhibited electrochemical properties prominently and for electrochemical purposes accordingly.

Peculiarities related to graphene as well as graphene oxide compared to allotropes of carbon were also discussed in detail. Aforesaid merits include such as excellent surface-area, high conductivity, great solubility, facile synthesis, and cheap source material as well. Though various technological advancements were explored yet space is available for improvement particularly for both electro-analytical and electrochemical sensors. Some of other electrochemical applications related to graphene oxide are still extendable covering further electrochemical applications towards future directions. Furthermore, critical challenges are still associated with such material as facile synthesis has been critically addressed. The structure of graphene oxide is also still incomplete at molecular level and therefore considered more important in literature. Other focus areas are supposed to be an understudy for further attention with respect to defects concerned with conductivity of graphene oxide. A brief understanding of electron flow on graphene oxide substrate/interface will also be an empty area of research available for further enhancement towards graphene oxide as well as other applications. Designs and approaches adopted, up till now, associated with manufacturing of graphene oxide devices are suggested critical in the future status of this material. Despite the aforementioned and highlighted challenges, graphene oxide applications associated with electrochemical sensors remain the key future application of graphene oxide.

Author details

Muhammad Ikram[1], Ali Raza[2], Khurram Shahzad[3], Ali Haider[4], Junaid Haider[5], Abdullah Khan Durrani[3], Asim Hassan Rizvi[3], Asghari Maqsood[6] and Mujtaba Ikram[3]*

1 Solar Cell Applications Research Lab, Department of Physics, Government College University Lahore, Punjab, Pakistan

2 Department of Physics, Riphah Institute of Computing and Applied Sciences (RICAS), Riphah International University, Lahore, Pakistan

3 Institute of Chemical Engineering and Technology (ICET), University of the Punjab (PU), Lahore, Punjab, Pakistan

4 Department of Clinical Medicine and Surgery, University of Veterinary and Animal Sciences, Lahore, Punjab, Pakistan

5 Tianjin Institute of Industrial Biotechnology, Chinese Academy of Sciences, Tianjin, China

6 Nanoscale Laboratory, Department of Physics, Air University, Islamabad, Pakistan

*Address all correspondence to: mujtaba.icet@pu.edu.pk

IntechOpen

© 2021 The Author(s). Licensee IntechOpen. This chapter is distributed under the terms of the Creative Commons Attribution License (http://creativecommons.org/licenses/by/3.0), which permits unrestricted use, distribution, and reproduction in any medium, provided the original work is properly cited.

References

[1] S. Ray, *Applications of graphene and graphene-oxide based nanomaterials*. William Andrew, 2015.

[2] W. Liu, S. Tan, Z. Yang, and G. Ji, "Hollow graphite spheres embedded in porous amorphous carbon matrices as lightweight and low-frequency microwave absorbing material through modulating dielectric loss," *Carbon,* vol. 138, pp. 143-153, 2018/11/01/2018.

[3] V. L. Deringer and G. Csányi, "Machine learning based interatomic potential for amorphous carbon," *Physical Review B,* vol. 95, no. 9, p. 094203, 03/03/2017.

[4] K. S. Novoselov *et al.*, "Electric Field Effect in Atomically Thin Carbon Films," *Science,* vol. 306, no. 5696, p. 666, 2004.

[5] A. K. Geim, "Graphene: Status and Prospects," *Science,* vol. 324, no. 5934, p. 1530, 2009.

[6] C. N. R. Rao, A. K. Sood, K. S. Subrahmanyam, and A. Govindaraj, "Graphene: The New Two-Dimensional Nanomaterial," vol. 48, no. 42, pp. 7752-7777, 2009.

[7] A. H. Castro Neto, F. Guinea, N. M. R. Peres, K. S. Novoselov, and A. K. Geim, "The electronic properties of graphene," *Reviews of Modern Physics,* vol. 81, no. 1, pp. 109-162, 01/14/ 2009.

[8] M. J. Allen, V. C. Tung, and R. B. Kaner, "Honeycomb Carbon: A Review of Graphene," *Chemical Reviews,* vol. 110, no. 1, pp. 132-145, 2010/01/13 2010.

[9] B. H. Hong, "Synthesis and applications of graphene for flexible electronics," in *69th Device Research Conference*, 2011, pp. 37-38.

[10] T. K. Das and S. Prusty, "Graphene-Based Polymer Composites and Their Applications," *Polymer-Plastics Technology and Engineering,* vol. 52, no. 4, pp. 319-331, 2013/03/16 2013.

[11] S. Stankovich *et al.*, "Graphene-based composite materials," *Nature,* vol. 442, no. 7100, pp. 282-286, 2006/07/01 2006.

[12] G. Eda and M. Chhowalla, "Graphene-based Composite Thin Films for Electronics," *Nano Letters,* vol. 9, no. 2, pp. 814-818, 2009/02/11 2009.

[13] T. Kuila *et al.*, "Preparation of functionalized graphene/linear low density polyethylene composites by a solution mixing method," *Carbon,* vol. 49, no. 3, pp. 1033-1037, 2011/03/01/ 2011.

[14] T. Kuila, S. Bose, A. K. Mishra, P. Khanra, N. H. Kim, and J. H. Lee, "Effect of functionalized graphene on the physical properties of linear low density polyethylene nanocomposites," *Polymer Testing,* vol. 31, no. 1, pp. 31-38, 2012/02/01/ 2012.

[15] A. J. Marsden *et al.*, "Electrical percolation in graphene–polymer composites," *2D Materials,* vol. 5, no. 3, p. 032003, 2018/06/01 2018.

[16] W. Gao *et al.*, "High-efficiency electromagnetic interference shielding realized in nacre-mimetic graphene/ polymer composite with extremely low graphene loading," *Carbon,* vol. 157, pp. 570-577, 2020/02/01/ 2020.

[17] X. Wang, W. Xing, P. Zhang, *L. song*, H. Yang, and Y. Hu, "Covalent functionalization of graphene with organosilane and its use as a reinforcement in epoxy composites," *Composites Science and Technology,* vol. 72, no. 6, pp. 737-743, 2012/03/27/ 2012.

[18] M. A. Rafiee, J. Rafiee, Z. Wang, H. Song, Z.-Z. Yu, and N. Koratkar,

"Enhanced Mechanical Properties of Nanocomposites at Low Graphene Content," *ACS Nano,* vol. 3, no. 12, pp. 3884-3890, 2009/12/22 2009.

[19] M. A. Rafiee *et al.*, "Fracture and Fatigue in Graphene Nanocomposites," vol. 6, no. 2, pp. 179-183, 2010.

[20] H. Bai, C. Li, and G. Shi, "Functional Composite Materials Based on Chemically Converted Graphene," vol. 23, no. 9, pp. 1089-1115, 2011.

[21] J. R. Potts, D. R. Dreyer, C. W. Bielawski, and R. S. Ruoff, "Graphene-based polymer nanocomposites," *Polymer,* vol. 52, no. 1, pp. 5-25, 2011/01/07/ 2011.

[22] N. H. Othman, M. Che Ismail, M. Mustapha, N. Sallih, K. E. Kee, and R. Ahmad Jaal, "Graphene-based polymer nanocomposites as barrier coatings for corrosion protection," *Progress in Organic Coatings,* vol. 135, pp. 82-99, 2019/10/01/ 2019.

[23] M. Gresil, Z. Wang, Q.-A. Poutrel, and C. Soutis, "Thermal Diffusivity Mapping of Graphene Based Polymer Nanocomposites," *Scientific Reports,* vol. 7, no. 1, p. 5536, 2017/07/17 2017.

[24] P. Govindaraj, B. Fox, P. Aitchison, and N. Hameed, "A Review on Graphene Polymer Nanocomposites in Harsh Operating Conditions," *Industrial & Engineering Chemistry Research,* vol. 58, no. 37, pp. 17106-17129, 2019/09/18 2019.

[25] J. Chen, B. Liu, and X. Gao, "Thermal properties of graphene-based polymer composite materials: A molecular dynamics study," *Results in Physics,* vol. 16, p. 102974, 2020/03/01/ 2020.

[26] A. Kausar, I. Rafique, and B. Muhammad, "Aerospace Application of Polymer Nanocomposite with Carbon Nanotube, Graphite, Graphene Oxide, and Nanoclay," *Polymer-Plastics Technology and Engineering,* vol. 56, no. 13, pp. 1438-1456, 2017/09/02 2017.

[27] D. Ponnamma *et al.*, "Graphene and graphitic derivative filled polymer composites as potential sensors," *Physical Chemistry Chemical Physics,* 10.1039/C4CP04418E vol. 17, no. 6, pp. 3954-3981, 2015.

[28] K. Qi, S.-y. Liu, and A. Zada, "Graphitic carbon nitride, a polymer photocatalyst," *Journal of the Taiwan Institute of Chemical Engineers,* vol. 109, pp. 111-123, 2020/04/01/ 2020.

[29] A. Kausar and S. Anwar, "Graphite Filler-Based Nanocomposites with Thermoplastic Polymers: A Review," *Polymer-Plastics Technology and Engineering,* vol. 57, no. 6, pp. 565-580, 2018/04/13 2018.

[30] A. Wang, C. Wang, L. Fu, W. Wong-Ng, and Y. Lan, "Recent Advances of Graphitic Carbon Nitride-Based Structures and Applications in Catalyst, Sensing, Imaging, and LEDs," *Nano-Micro Letters,* vol. 9, no. 4, p. 47, 2017/06/08 2017.

[31] D. R. Dreyer, S. Murali, Y. Zhu, R. S. Ruoff, and C. W. Bielawski, "Reduction of graphite oxide using alcohols," *Journal of Materials Chemistry,* 10.1039/C0JM02704A vol. 21, no. 10, pp. 3443-3447, 2011.

[32] Q. Wu, Y. Xu, Z. Yao, A. Liu, and G. Shi, "Supercapacitors Based on Flexible Graphene/Polyaniline Nanofiber Composite Films," *ACS Nano,* vol. 4, no. 4, pp. 1963-1970, 2010/04/27 2010.

[33] H. Pang, Y.-C. Zhang, T. Chen, B.-Q. Zeng, and Z.-M. Li, "Tunable positive temperature coefficient of resistivity in an electrically conducting polymer/graphene composite," vol. 96, no. 25, p. 251907, 2010.

[34] C. Li *et al.*, "Graphene Nano-"patches" on a Carbon

Nanotube Network for Highly Transparent/Conductive Thin Film Applications," *The Journal of Physical Chemistry C,* vol. 114, no. 33, pp. 14008-14012, 2010/08/26 2010.

[35] J. Lee, P. Lee, H. Lee, D. Lee, S. S. Lee, and S. H. Ko, "Very long Ag nanowire synthesis and its application in a highly transparent, conductive and flexible metal electrode touch panel," *Nanoscale,* 10.1039/C2NR31254A vol. 4, no. 20, pp. 6408-6414, 2012.

[36] H. Kim and C. W. Macosko, "Processing-property relationships of polycarbonate/graphene composites," *Polymer,* vol. 50, no. 15, pp. 3797-3809, 2009/07/17/ 2009.

[37] M. H. Yang, B. G. Choi, H. Park, W. H. Hong, S. Y. Lee, and T. J. Park, "Development of a Glucose Biosensor Using Advanced Electrode Modified by Nanohybrid Composing Chemically Modified Graphene and Ionic Liquid," *Electroanalysis,* https://doi.org/10.1002/elan.200900556 vol. 22, no. 11, pp. 1223-1228, 2010/06/01 2010.

[38] X. Kang, J. Wang, H. Wu, I. A. Aksay, J. Liu, and Y. Lin, "Glucose Oxidase–graphene–chitosan modified electrode for direct electrochemistry and glucose sensing," *Biosensors and Bioelectronics,* vol. 25, no. 4, pp. 901-905, 2009/12/15/ 2009.

[39] L. Xiao *et al.*, "Flexible, Stretchable, Transparent Carbon Nanotube Thin Film Loudspeakers," *Nano Letters,* vol. 8, no. 12, pp. 4539-4545, 2008/12/10 2008.

[40] W. Zhang *et al.*, "Tribological properties of oleic acid-modified graphene as lubricant oil additives," *Journal of Physics D: Applied Physics,* vol. 44, no. 20, p. 205303, 2011/05/03 2011.

[41] J. Lin, L. Wang, and G. Chen, "Modification of Graphene Platelets and their Tribological Properties as a Lubricant Additive," *Tribology Letters,* vol. 41, no. 1, pp. 209-215, 2011/01/01 2011.

[42] S. Choudhary, H. P. Mungse, and O. P. Khatri, "Dispersion of alkylated graphene in organic solvents and its potential for lubrication applications," *Journal of Materials Chemistry,* 10.1039/C2JM34741E vol. 22, no. 39, pp. 21032-21039, 2012.

[43] T. T. Baby and S. Ramaprabhu, "Investigation of thermal and electrical conductivity of graphene based nanofluids," vol. 108, no. 12, p. 124308, 2010.

[44] T. T. Baby and S. Ramaprabhu, "Enhanced convective heat transfer using graphene dispersed nanofluids," *Nanoscale Research Letters,* vol. 6, no. 1, p. 289, 2011/04/04 2011.

[45] T. Kuila, P. Banerjee, N. C. J. A. C. M. Murmu, and Technology, "Surface Modification of Graphene," p. 35, 2013.

[46] G. Eda, G. Fanchini, and M. Chhowalla, "Large-area ultrathin films of reduced graphene oxide as a transparent and flexible electronic material," *Nature Nanotechnology,* vol. 3, no. 5, pp. 270-274, 2008/05/01 2008.

[47] K. Rana, J. Singh, and J.-H. Ahn, "A graphene-based transparent electrode for use in flexible optoelectronic devices," *Journal of Materials Chemistry C,* 10.1039/C3TC32264E vol. 2, no. 15, pp. 2646-2656, 2014.

[48] S. Chun *et al.*, "Single-Layer Graphene-Based Transparent and Flexible Multifunctional Electronics for Self-Charging Power and Touch-Sensing Systems," *ACS Applied Materials & Interfaces,* vol. 11, no. 9, pp. 9301-9308, 2019/03/06 2019.

[49] K. S. Kim *et al.*, "Large-scale pattern growth of graphene films for stretchable transparent electrodes," *Nature,* vol. 457, no. 7230, pp. 706-710, 2009/02/01 2009.

[50] Y. Ma and L. Zhi, "Graphene-Based Transparent Conductive Films: Material Systems, Preparation and Applications," vol. 3, no. 1, p. 1800199, 2019.

[51] F. Bonaccorso, Z. Sun, T. Hasan, and A. C. Ferrari, "Graphene photonics and optoelectronics," *Nature Photonics,* vol. 4, no. 9, pp. 611-622, 2010/09/01 2010.

[52] J. Park *et al.*, "Work-Function Engineering of Graphene Electrodes by Self-Assembled Monolayers for High-Performance Organic Field-Effect Transistors," *The Journal of Physical Chemistry Letters,* vol. 2, no. 8, pp. 841-845, 2011/04/21 2011.

[53] K. Kim *et al.*, "Ultrathin Organic Solar Cells with Graphene Doped by Ferroelectric Polarization," *ACS Applied Materials & Interfaces,* vol. 6, no. 5, pp. 3299-3304, 2014/03/12 2014.

[54] K. Sint, B. Wang, and P. Král, "Selective Ion Passage through Functionalized Graphene Nanopores," *Journal of the American Chemical Society,* vol. 130, no. 49, pp. 16448-16449, 2008/12/10 2008.

[55] D.-e. Jiang, V. R. Cooper, and S. Dai, "Porous Graphene as the Ultimate Membrane for Gas Separation," *Nano Letters,* vol. 9, no. 12, pp. 4019-4024, 2009/12/09 2009.

[56] M. Shan *et al.*, "Influence of chemical functionalization on the CO2/N2 separation performance of porous graphene membranes," *Nanoscale,* 10.1039/C2NR31402A vol. 4, no. 17, pp. 5477-5482, 2012.

[57] A. Ali, R. Pothu, S. H. Siyal, S. Phulpoto, M. Sajjad, and K. H. Thebo, "Graphene-based membranes for CO2 separation," *Materials Science for Energy Technologies,* vol. 2, no. 1, pp. 83-88, 2019/04/01/ 2019.

[58] D. Cohen-Tanugi and J. C. Grossman, "Water Desalination across Nanoporous Graphene," *Nano Letters,* vol. 12, no. 7, pp. 3602-3608, 2012/07/11 2012.

[59] F. Zhou, M. Fathizadeh, and M. Yu, "Single- to Few-Layered, Graphene-Based Separation Membranes," vol. 9, no. 1, pp. 17-39, 2018.

[60] A. Raza *et al.*, "Enhanced industrial dye degradation using Co doped in chemically exfoliated MoS2 nanosheets," *Applied Nanoscience,* vol. 10, no. 5, pp. 1535-1544, 2019.

[61] S. C. O'Hern *et al.*, "Selective Molecular Transport through Intrinsic Defects in a Single Layer of CVD Graphene," *ACS Nano,* vol. 6, no. 11, pp. 10130-10138, 2012/11/27 2012.

[62] N. Song, X. Gao, Z. Ma, X. Wang, Y. Wei, and C. Gao, "A review of graphene-based separation membrane: Materials, characteristics, preparation and applications," *Desalination,* vol. 437, pp. 59-72, 2018/07/01/ 2018.

[63] L. Nie, C. Y. Chuah, T.-H. Bae, and J.-M. Lee, "Graphene-Based Advanced Membrane Applications in Organic Solvent Nanofiltration," vol. n/a, no. n/a, p. 2006949.

[64] A. Raza *et al.*, "A comparative study of dirac 2D materials, TMDCs and 2D insulators with regard to their structures and photocatalytic/sonophotocatalytic behavior," *Applied Nanoscience,* 2020.

[65] M. Ikram, M. I. Khan, A. Raza, M. Imran, A. Ul-Hamid, and S. Ali, "Outstanding performance of silver-decorated MoS2 nanopetals used as nanocatalyst for synthetic dye degradation," *Physica E: Low-dimensional Systems and Nanostructures,* vol. 124, 2020.

[66] S. P. Koenig, L. Wang, J. Pellegrino, and J. S. Bunch, "Selective molecular sieving through porous graphene,"

Nature Nanotechnology, vol. 7, no. 11, pp. 728-732, 2012/11/01 2012.

[67] L. Wang *et al.*, "Molecular valves for controlling gas phase transport made from discrete ångström-sized pores in graphene," *Nature Nanotechnology,* vol. 10, no. 9, pp. 785-790, 2015/09/01 2015.

[68] J.-J. Lu, Y.-H. Gu, Y. Chen, X. Yan, Y.-J. Guo, and W.-Z. Lang, "Ultrahigh permeability of graphene-based membranes by adjusting D-spacing with poly (ethylene imine) for the separation of dye wastewater," *Separation and Purification Technology,* vol. 210, pp. 737-745, 2019/02/08/ 2019.

[69] M. Ikram, A. Raza, M. Imran, A. Ul-Hamid, A. Shahbaz, and S. Ali, "Hydrothermal Synthesis of Silver Decorated Reduced Graphene Oxide (rGO) Nanoflakes with Effective Photocatalytic Activity for Wastewater Treatment," *Nanoscale Res Lett,* vol. 15, no. 1, p. 95, Apr 28 2020.

[70] C. Casiraghi *et al.*, "Rayleigh Imaging of Graphene and Graphene Layers," *Nano Letters,* vol. 7, no. 9, pp. 2711-2717, 2007/09/01 2007.

[71] S. Park, K.-S. Lee, G. Bozoklu, W. Cai, S. T. Nguyen, and R. S. Ruoff, "Graphene Oxide Papers Modified by Divalent Ions—Enhancing Mechanical Properties via Chemical Cross-Linking," *ACS Nano,* vol. 2, no. 3, pp. 572-578, 2008/03/25 2008.

[72] D. Li, M. B. Müller, S. Gilje, R. B. Kaner, and G. G. Wallace, "Processable aqueous dispersions of graphene nanosheets," *Nature Nanotechnology,* vol. 3, no. 2, pp. 101-105, 2008/02/01 2008.

[73] Y. Xu, H. Bai, G. Lu, C. Li, and G. Shi, "Flexible Graphene Films via the Filtration of Water-Soluble Noncovalent Functionalized Graphene Sheets," *Journal of the American Chemical Society,* vol. 130, no. 18, pp. 5856-5857, 2008/05/01 2008.

[74] Y. H. Wu, T. Yu, and Z. X. Shen, "Two-dimensional carbon nanostructures: Fundamental properties, synthesis, characterization, and potential applications," vol. 108, no. 7, p. 071301, 2010.

[75] Z. Chen, L. Jin, W. Hao, W. Ren, and H. M. Cheng, "Synthesis and applications of three-dimensional graphene network structures," *Materials Today Nano,* vol. 5, p. 100027, 2019/03/01/ 2019.

[76] A. A. Green and M. C. Hersam, "Emerging Methods for Producing Monodisperse Graphene Dispersions," *The Journal of Physical Chemistry Letters,* vol. 1, no. 2, pp. 544-549, 2010/01/21 2010.

[77] K. Balasubramanian and M. Burghard, "Biosensors based on carbon nanotubes," *Analytical and Bioanalytical Chemistry,* vol. 385, no. 3, pp. 452-468, 2006/06/01 2006.

[78] S. H. Güler, Ö. Güler, and E. Evin, "The production of graphene nano layers by using milling—exfoliation hybrid process," *Fullerenes, Nanotubes and Carbon Nanostructures,* vol. 25, no. 1, pp. 34-39, 2017/01/02 2017.

[79] V. C. Tung, M. J. Allen, Y. Yang, and R. B. Kaner, "High-throughput solution processing of large-scale graphene," *Nature Nanotechnology,* vol. 4, no. 1, pp. 25-29, 2009/01/01 2009.

[80] A. A. Green and M. C. Hersam, "Solution Phase Production of Graphene with Controlled Thickness via Density Differentiation," *Nano Letters,* vol. 9, no. 12, pp. 4031-4036, 2009/12/09 2009.

[81] J. Cai *et al.*, "Atomically precise bottom-up fabrication of graphene nanoribbons," *Nature,* vol. 466, no. 7305, pp. 470-473, 2010/07/01 2010.

[82] N. Shaari and S. K. Kamarudin, "Graphene in electrocatalyst and proton

conductiong membrane in fuel cell applications: An overview," *Renewable and Sustainable Energy Reviews,* vol. 69, pp. 862-870, 2017/03/01/ 2017.

[83] R. P. Pandey, G. Shukla, M. Manohar, and V. K. Shahi, "Graphene oxide based nanohybrid proton exchange membranes for fuel cell applications: An overview," *Advances in Colloid and Interface Science,* vol. 240, pp. 15-30, 2017/02/01/ 2017.

[84] J. Li *et al.*, "Non-destructive modification on Nafion membrane via in-situ inserting of sheared graphene oxide for direct methanol fuel cell applications," *Electrochimica Acta,* vol. 282, pp. 362-368, 2018/08/20/ 2018.

[85] Z.-L. Wang, D. Xu, J.-J. Xu, and X.-B. Zhang, "Oxygen electrocatalysts in metal–air batteries: from aqueous to nonaqueous electrolytes," *Chemical Society Reviews,* 10.1039/C3CS60248F vol. 43, no. 22, pp. 7746-7786, 2014.

[86] C. Zhang *et al.*, "Single-Atomic Ruthenium Catalytic Sites on Nitrogen-Doped Graphene for Oxygen Reduction Reaction in Acidic Medium," *ACS Nano,* vol. 11, no. 7, pp. 6930-6941, 2017/07/25 2017.

[87] H. Wang, Y. Liang, Y. Li, and H. Dai, "Co1−xS–Graphene Hybrid: A High-Performance Metal Chalcogenide Electrocatalyst for Oxygen Reduction," vol. 50, no. 46, pp. 10969-10972, 2011.

[88] S. Guo and S. Sun, "FePt Nanoparticles Assembled on Graphene as Enhanced Catalyst for Oxygen Reduction Reaction," *Journal of the American Chemical Society,* vol. 134, no. 5, pp. 2492-2495, 2012/02/08 2012.

[89] Y. Liang *et al.*, "Covalent Hybrid of Spinel Manganese–Cobalt Oxide and Graphene as Advanced Oxygen Reduction Electrocatalysts," *Journal of the American Chemical Society,* vol. 134, no. 7, pp. 3517-3523, 2012/02/22 2012.

[90] M. A. Abdelkareem, T. Yoshitoshi, T. Tsujiguchi, and N. Nakagawa, "Vertical operation of passive direct methanol fuel cell employing a porous carbon plate," *Journal of Power Sources,* vol. 195, no. 7, pp. 1821-1828, 2010/04/02/ 2010.

[91] J. C. M. da Silva, T. Lopes, A. O. Neto, and E. V. Spinacé, "Electrocatalysts for direct ethanol fuel cells," 2019.

[92] Y. Ito, T. Takeuchi, T. Tsujiguchi, M. A. Abdelkareem, and N. Nakagawa, "Ultrahigh methanol electro-oxidation activity of PtRu nanoparticles prepared on TiO2-embedded carbon nanofiber support," *Journal of Power Sources,* vol. 242, pp. 280-288, 2013/11/15/ 2013.

[93] E. T. Sayed *et al.*, "Direct urea fuel cells: Challenges and opportunities," *Journal of Power Sources,* vol. 417, pp. 159-175, 2019/03/31/ 2019.

[94] Y. Shao, J.-P. Dodelet, G. Wu, and P. Zelenay, "PGM-Free Cathode Catalysts for PEM Fuel Cells: A Mini-Review on Stability Challenges," vol. 31, no. 31, p. 1807615, 2019.

[95] A. Borenstein, O. Hanna, R. Attias, *S. luski*, T. Brousse, and D. Aurbach, "Carbon-based composite materials for supercapacitor electrodes: a review," *Journal of Materials Chemistry A,* 10.1039/C7TA00863E vol. 5, no. 25, pp. 12653-12672, 2017.

[96] S. I. Wong, H. Lin, J. Sunarso, B. T. Wong, and B. Jia, "Optimization of ionic-liquid based electrolyte concentration for high-energy density graphene supercapacitors," *Applied Materials Today,* vol. 18, p. 100522, 2020/03/01/ 2020.

[97] S. Korkmaz and İ. A. Kariper, "Graphene and graphene oxide based aerogels: Synthesis, characteristics and supercapacitor applications," *Journal*

of Energy Storage, vol. 27, p. 101038, 2020/02/01/ 2020.

[98] D. Chen, L. Tang, and J. Li, "Graphene-based materials in electrochemistry," *Chemical Society Reviews,* 10.1039/B923596E vol. 39, no. 8, pp. 3157-3180, 2010.

[99] A. S. ARICÒ, P. BRUCE, B. SCROSATI, J.-M. TARASCON, and W. V. SCHALKWIJK, "Nanostructured materials for advanced energy conversion and storage devices," in *Materials for Sustainable Energy*, pp. 148-159.

Section 2

Sustainable Fabrication of Advanced Carbon Materials for Applied Applications

Chapter 3

Carbon Nanotubes Integrated Hydroxyapatite Nano-Composite for Orthopaedic and Tissue Engineering Applications

Khalid Parwez, Arun A. Bhagwath, Asif Zawed, Bhagwan Rekadwad and Suman V. Budihal

Abstract

The reassessment of the literature stipulates that an increasing amount of research in exploring the Hydroxyapatite Carbon Nanotubes (HA-CNT) system for orthopedic application. Chemical precipitation, CNT functionalization, and spray drying are the routinely used methods for CNT dispersal in HA matrix for the application such as bone tissue engineering, nanostructured scaffolds, dental regeneration, myocardial regeneration, and skin regeneration. Although mechanical strength and biocompatibility is a substantial concern for the fabrication of structures. Developing composite and bioceramic scaffolding with different natural and synthetic biomaterials are the futuristic approach in the biomedical engineering field. The problems such as biocompatibility, biodegradability, and mechanical resistance can be solved by combining natural, and artificial biomaterials. The natural biomaterials, such as collagen, cellulose, chitosan, have a close resemblance to the natural extracellular matrix (ECM). These materials are biocompatible, biodegradable. The artificial biomaterials, such as Poly Vinyl Pyrrolidone (PVP), Poly Capro Lactone (PCL), Poly Ethylene Glycol (PEG), and Poly Lactic Acid (PLA) are also the material of choice for the fabrication of the composite materials. Additional effort is necessary to fabricate biocompatible composite scaffolding for tissue engineering. Moreover, vascularization, differentiation, cellular proliferation, and cells to scaffold interaction are the foremost challenges in the area of tissue engineering that remains to overcome.

Keywords: Nanocomposites, Carbon Nanomaterial, Fabrication, Electrospinning, Bone Tissue Engineering

1. Introduction

The bone defects are generally caused by trauma, tumor resection, deformity, and infections. This kind of large bone defect has increased in recent years. Lack of resources for bone grafting as well as rejections of the grafts are potential obstacles in the treatment of these defects. Tissue engineering is considered the emerging viable technology for reconstructing tissues and organs structure. Tissue engineering is also helpful in repair, and reinforce the functions of dented tissues.

The electrospinning techniques have been widely used for fabricating ultrafine, and unremitting nano-fibres. The bone and dentin are natural mineralized hard tissues with extremely compound and classified structures [1–3]. The3-dimensional structures to the hard mineralized tissue are produced by the deposition of minerals to the structured organic matrix [4]. The primary minerals of the tissues are hydroxyapatite (HAp) (Ca10(PO4)6(OH)2) found in bone and dentine [5, 6].

To fabricate the bone and dentin like biomaterials, it is essential that the scientist from bioengineering, biomedical, biology, chemistry, and materials science should come together. The multidisciplinary approach of fabrication shall create biocompatible hard tissue with improved mechanical properties. The large aspect ratio of nanomaterials has attracted the extensive attention of material scientists across the globe. The large aspect ratio enables the nanomaterials to have a large surface area in less volume.

The following points shall be considered for the development of biomaterials for hard tissue regeneration.

i. The material shall be compatible with the regenerated tissue.

ii. The time for tissue regeneration and healing shall be reduced.

iii. There shall not be any inflammatory or toxic response to the host because of the materials used.

iv. The mechanical properties of material shall be equivalent to the natural bone or dentine material.

Recently, the biomimetic approaches to synthesize the natural composite materials are being discussed by the scientific communities [5]. The biomimetic approach of fabrication enables the nanomaterials to spread to a specific orientation in the matrix. Hence, it is expected that the newly synthesize composite shall have the morphology and composition of the natural tissues. The PVA is a water-soluble synthetic polymer. The PVA backbone is exceedingly interconnected by H-bond. The presence of the abundant amount of hydroxyl groups (–OH) in PVA polymer enables them to have better mechanical, chemical, and biological properties e.g., better tensile strength, excellent chemical resistance, biocompatibility, and stealth resistance. Moreover, the –OH group also allows appending the biological molecules like antibodies, DNA collagens, and hyaluronidase to PVA without any compatibility issues [6, 7]. The biocompatibility and biodegradability of PVA polymer to living tissues results in a wide range of biomedical and biopharmaceutical applications [8–10]. Two reasons to select PVA/HA nanocomposites for this study are (a) PVA and HA both has abundant hydroxyl groups (–OH), so, PVA can strongly interact with HA and, (b) aqueous phase electrospinning of PVA/HA nanocomposite is possible.

We propose an alternative approach to PVA-HA nanocomposite manufacturing in this research, which shall have a promising future application related to hard-tissue engineering. The electrospun PVA-HA composite shall be an alternative to bone replacement and dentine coating. Because, the electrospun composite shall have similar physicochemical properties, and resemble the nanostructure of living bone.

The synthetic hydroxyapatite is biocompatible, bioactive, and osteoconductive material [11], which is similar to the hydroxyapatite present in the living system. The chemical composition and molecular structures of the synthetic hydroxyapatite are similar to the minerals present in the dentine, cartilage, and bone. Thus, the synthetic HA has been attracting the attention of the scientist in biomedical an application such as coating for the implant, prosthesis, and replacement of defective bone [12].

Hydroxyapatite is the integral component of natural bone, which is bioactive, biocompatible, and osteoconductive. Thus, this material is most widely used in tissue engineering. The brittleness and low biodegradability are considered as the limiting factor for this material [13, 14]. Hence, making the composite with PVA can enhance mechanical strength and biodegradability. The PVA/HA nanocomposite shall have a suitable pore size, good aspect ratio, and biocompatibility to the living cells [15].

Carbon nanotubes (CNTs) are classified as single-walled carbon nanotubes (SWNTs) and multi-walled carbon nanotubes (MWNTs). The CNTs are considered as a suitable nanomaterial for biomedical application because of their aspect ratio, excellent mechanical, thermal, and electromagnetic properties [16]. CNTs are proved to be biocompatible and promote attachments and proliferation of osteoblast cells. Hence, it can be assumed that the presence of CNTs in the matrix shall result in better and improved composite fabrication.

The electrospinning is a technology introduced in the year 1934. This technology is successful in producing nanofibers at a very low cost [17]. The cost-effectiveness and the dimension of produced fibers enable this technology to be used in the field of tissue engineering [18–20]. The as-produced structure of HA by electrospinning resembles natural ECM, which enables them to be used as a scaffold for tissue regeneration, immobilizing enzymes, wound dressing materials, and tissue culture [21–24]. In the present study, we synthesized composite fibers using PVA and HA via electrospinning and characterized by UTM, SEM analysis. The electro-spun three-dimensional structure is microscopically similar to bone.

2. Review of literature

Miniaturization of the devices encouraged the scientist to explore a polymer nanofiber. The electrospinning techniques enable them to produce controlled fiber diameter with high strength, which couldn't be achieved by the conventional fiber processing techniques.

The microstructure formed during electrospinning leads to the fabrication of strong fibers by systematic arrangements. It is important to understand the effect of electrospinning parameters on morphology, tensile strength, and fiber diameter. The electrospinning parameters affect the inherent structure and overall deformation behavior. The deformation behavior, tensile strength, and fiber size are mostly affected by crystallinity and molecular orientation. Whereas, the electro wiring affects the fiber diameters during the process of electrospinning [25–29]. Furthermore, the electric field applied during the electrospinning encourages the self-assembly of the fillers, which leads to composite synthesis in spatial arrangements. The CNTs and carbon black (CB) are the most commonly used fillers in the electrospinning process. These fillers disperse themselves within the fiber and provide high strength and toughness to the composite. This simulating behavior of the fillers make them a good candidate for the application such as tissue engineering, filtration, and advance nanofibers [30–34].

The characterization of the nanofibers for the traction behavior is the challenging part of the field because the nanofibers get deformed with low load. Consequently, the mechanical integrity of the nanofibers is less studied and understood. The characterization of the mechanical deformation of the nonwoven fabrics prepared by electrospinning is the need of the time. There are several parameters that affect the tensile properties of the nonwoven fibers, such as, entanglement, orientation, porosity, fiber-fiber interaction, and size distribution of the fibers. Although, these parameters are not readily controllable in the nonwoven fibers, hence, characterization of the tensile properties of the single fibers is of very much interest. In recent studies, it has been shown that the cutting effect and size effects are important to control the intrinsic structure and diameter respectively [35–37].

3. Materials and Methods

3.1 Materials

- PVA (Polyvinyl chloride).
- HA (Hydroxy apatite).
- CNT (Carbon nanotubes).
- 0.5 M, 5 ml of Calcium Nitrate [$Ca(NO_3)_2 \cdot 4H_2O$].
- 0.5 M, 10 ml of Diammonium hydrogen phosphate [$(NH_4)_2HPO_4$].
- Ethanol.
- Distilled water.
- Magnetic stirrer.
- Cotton.
- Beakers.

3.2 Apparatus required

- Electrospinning Machine.
- Sonicator.
- Stirrer machine.
- UTM (Universal testing machine).
- SEM (Scanning electron microscopy).

4. Methods

4.1 PVA film formations

- 18 gm of PVA is taken in a beaker and It is mixed with 182 ml of distilled water.
- 200 ml of PVA solution is continuously stirred for 3 hrs using magnetic stirrer.

- Then this solution is loaded in a syringe upto 2 ml and fixed in the syringe clamp.
- Then either the drum collector or the square collector is connected so that the film is formed on it.
- Few parameters are set before starting the machine which are as followed:

    i. Voltage - 12Kv.

    ii. Flow-rate – 0.2 ml/hr.

    iii. Syringe volume -2 ml.

    iv. Needle diameter – 0.5 mm.

    v. Needle length -25 mm.

    vi. Distance between collector & syringe -18 cm.

    vii. Time of run – 8 hrs.

4.2 Formation of HA–PVA film

- 15 ml of PVA is taken in a beaker.
- 0.59gm of Ca(NO3)2 4H2O is added to 5 ml of distilled water.
- 0.57gm of (NH4)2HPO4 is added to 10 ml of distilled water.
- pH should be adjusted more than 10.
- The solution is stirred and ammonia solution is added into it.
- After preparing the solution it is loaded into the syringe upto 2 ml and make sure there won't be any bubble formation in the syringe.
- Then the syringe is adjusted to the suitable position.
- After that few parameters are set in order to form the film which are as follows:

    i. Voltage – 19Kv.

    ii. Flow - rate – 0.2 ml/hr.

    iii. Syringe volume – 2 ml.

    iv. Needle diameter – 1.2 mm.

    v. Needle length –38 mm.

    vi. Distance between the collector & syringe –

vii. Time of run – 8 hrs.

viii. Drum collector speed – 2000 rpm.

4.3 Formation of PVA-CNT film

- 15 ml of PVA is taken in a beaker and 1% of CNT (0.15gm) and sonicate it.
- Prepare two solutions of same quantity and sonicate them for 3 hrs.
- After that syringe is loaded with the solution which is prepared upto 2 ml.
- Then it is adjusted appropriately.
- Drum collector is connected on which the film will be formed.
- Few parameters are adjusted before starting the machine which are as follows:

i. Voltage – 19Kv.

ii. Flow-rate – 0.2 ml/hr.

iii. Syringe volume – 2 ml.

iv. Needle diameter – 1.2 mm.

v. Drum speed – 2000 rpm.

5. Characterization

5.1 Surface topography characterization

The surface morphology and atomic structure of PVA, PVA/HA, PVA/CNT (**Figure 1**) scaffolds were examined under the FESEM (Hitachi SU6600, Japan) and EDS (Horiba-EMAX). Prior to the microstructure analysis, specimens were coated with gold using an ion sputter coating instrument (Hitachi E – 1010, Japan) with a current set at a 15 mA for a coating time of 15 s.

5.2 Universal testing machine (UTM) analysis

A universal testing machine (UTM), also known as a universal tester materials testing machine or materials test frame, is used to test the tensile stress and compressive strength of materials. It is named after the fact that it can perform many standard tensile and compression test on materials, components and structures.

6. Results and discussion

6.1 Scanning electron microscope (SEM)

SEM micrographs of the samples were taken using (Zeiss) electron microscope, Japan. The samples were coated with gold using sputtering technique. Scanning

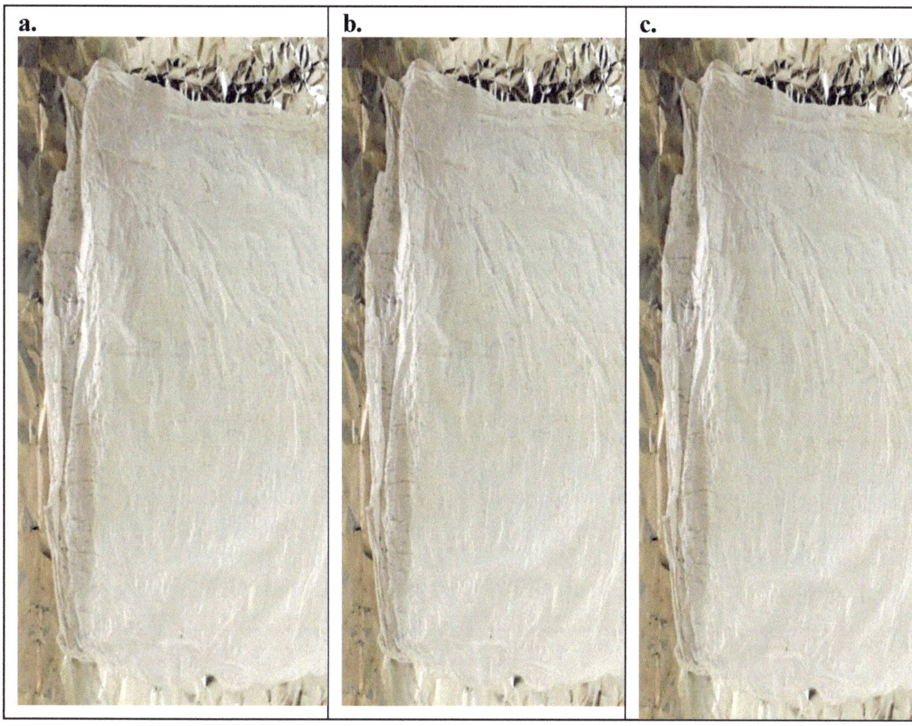

Figure 1.
a. PVA Film, b. PVA-HA Film, c. PVA-CNT Film prepared through Electrospinning.

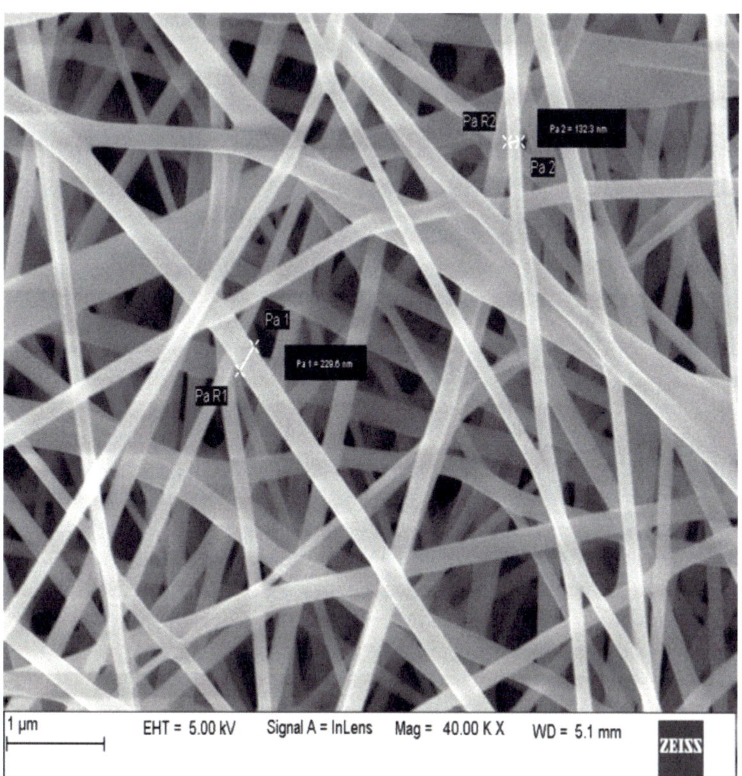

Figure 2.
SEM image of the PVA film, scale 1 μm and 4K X magnification.

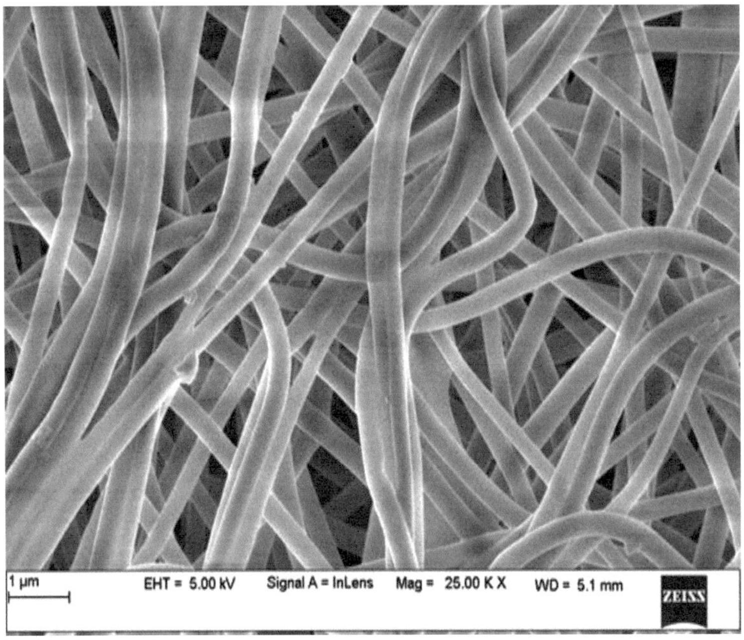

Figure 3.
SEM image of the PVA film, scale 1 μm and 40K X magnification.

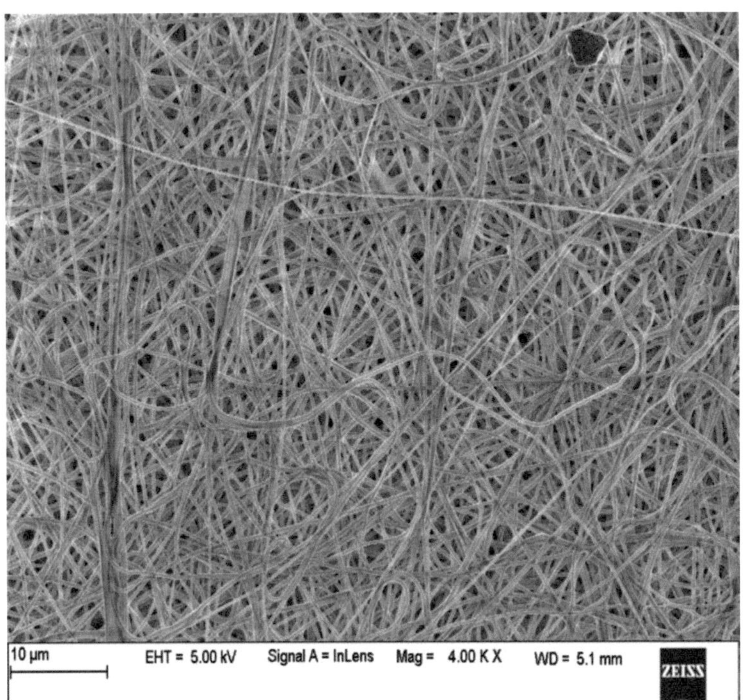

Figure 4.
SEM image of the PVA-HA film, scale 10 μm and 4K X magnification.

electron microscopy (SEM) was used to investigate the morphology of different types of films. To analyze complex structural variations upon undergoing electrospinning, and the interactions between HA and PVA. Due to abundant -OH group

Sample	Area (mm^2)	Tensile strength (MPA)	Young Modulus (GPA)	Strain at break (MPA)	Stress at break (MPA)	Percentage total elongation at fracture
Trial 1	0.42	6.06311	29.12233	0.355376	2.018833	35.5378
Trial 2	0.14	24.954	217.310	0.960583	12.66005	96.0584

Table 1.
UTM analysis for PVA films.

present in PVA and HA nanorods, they strongly interact via H-bonding within the electrospun HA-PVA nanocomposite fibres, improving thermal property.

In PVA film, at a scale of 10 μm, EHT = 5.00 kV and 4 K X magnification, the PVA film appeared like thin fibres arranged very closely (**Figure 2**). By Increasing the magnification to 40 K X and decreasing the scale to 1 μm, we observed combination of fibres with different sizes (**Figure 3**) ranging from 130 to 200 nm. In PVA-HA Film, at a scale of 10 μm, EHT = 5.00 kV and 4.00 K X magnification, the fibres are densely populated having an average size of 300 nm. Structure of nanofibres exist with relatively uniform dispersion with PVA.

After the successful binding of PVA with HA, carbon nanotubes were added with PVA to increase the strength of the film and give stability to the unwoven fibres. Since it has remarkable mechanical, thermal properties and outstanding electrical conductivity. It is much stronger and more functional fibres than current traditional industrial fibres. Hence in PVA-CNT film SEM gives successful result producing nano fibres, at a scale of 10 μm, EHT = 5.00 kV and 4.00 K X magnification (**Figure 4**) by increasing the magnification the fibres can be observed clearly.

6.2 Universal testing machine (UTM) analysis

Test is carried out to obtain the stress and strain at break and determine tensile strength of the fibres. This test is useful in getting the information of mechanical behavior and the engineering performance of the material. The first trial carried out on the film showed reduced tensile strength of 6.0631 MPA with young's modulus of 29.122 GPA. But the trial gave an improved result with increased tensile strength of 24.954 MPA and 217.310 GPA.

The percentage of elongation at fracture was also seen increasing from trial 1 to trial 2 from 35.5378 to 96.0584 respectively **Table 1**.

7. Conclusions

We prepared HA/PVA/CNT nanofibers nanocomposite via in situ co-precipitation method through electrospinning process. The produced nanofibers are a biomimetic composite of 200-300 nm. UTM analysis of the produced fibers proved to have better young modulus and tensile strength. The nano-bio materials are supposed to stimulate cell adhesion to the ECM matrix, and subsequent mineralization and tissue regeneration. The electrospun HA/PVA/CNT nanofibers shall have the potential for bone tissue engineering.

A nano-biomaterials-based scaffold shall have a high aspect ratio, better wettability, improved mechanical properties. These properties of nano-biomaterial-based scaffold make them desirable for tissue engineering. The scaffold-based on HA/PVA/CNT nanofibers are expected to have better regeneration capacity in vivo, compared to the conventional counterpart. Bone tissue engineering research requires the combination of mechanically strong biocompatible materials, which is essential in cell adhesion and proliferation. Hence, we propose the combination of these materials with stem cells and various growth factors as the future strategies in the field of biomedical and bone tissue engineering. We propose the two tools (i) nano-scaffoldings, and (ii) cell micro-encapsulation for equal distribution of the cells within the scaffolding. Furthermore, this research shall be validated by in vitro & in vivo studies to establish the efficacy of the tailor-made nano-bio materials for orthopedic and craniofacial applications.

Author details

Khalid Parwez[1]*, Arun A. Bhagwath[1], Asif Zawed[2], Bhagwan Rekadwad[1] and Suman V. Budihal[3]

1 Yenepoya Research Centre, Yenepoya (Deemed to be University), Deralakatte, Mangalore, Karnataka, India

2 Indian Pharmacopoeia Commission, Ministry of Health and Family Welfare, Government of India, Gaziabad, U.P., India

3 Department of Physiology, Kasturba Medical College, Mangalore, Manipal Academy of Higher Education (MAHE), Manipal, Karnataka, India

*Address all correspondence to: drkhalidpm@gmail.com

IntechOpen

© 2021 The Author(s). Licensee IntechOpen. This chapter is distributed under the terms of the Creative Commons Attribution License (http://creativecommons.org/licenses/by/3.0), which permits unrestricted use, distribution, and reproduction in any medium, provided the original work is properly cited.

References

[1] Weiner S, Traub W. Bone structure: from ångstroms to microns. The FASEB journal. 1992 Feb;6(3):879-85.

[2] Rho JY, Kuhn-Spearing L, Zioupos P. Mechanical properties and the hierarchical structure of bone. Medical engineering & physics. 1998 Mar 1;20(2):92-102.

[3] Fratzl P. Characterizing natural fibre composites with hierarchical structure. Fibre Diffraction Rev. 2002 Apr;210:31-9.

[4] Park J B and Lakes R S 1992 Biomaterials 2nd edition (New York: Plenum) p 194

[5] Ravaglioli A, Krajewski 'Bioceramics A. materials properties applications'.

[6] Elliott JC. Handbook of Structure and Chemistry of the Apatite and Other Calcium Orthophosphates. Vol. 18.

[7] Butler WT, Ritchie HE. The nature and functional significance of dentin extracellular matrix proteins. International Journal of Developmental Biology. 2003 Feb 1;39(1):169-79.

[8] Park J, Lutz R, Felszeghy E, Wiltfang J, Nkenke E, Neukam FW, Schlegel KA. The effect on bone regeneration of a liposomal vector to deliver BMP-2 gene to bone grafts in peri-implant bone defects. Biomaterials. 2007 Jun 1;28(17):2772-82.

[9] Kocialkowski A, Wallace WA, Prince HG. Clinical experience with a new artificial bone graft: preliminary results of a prospective study. Injury. 1990 May 1;21(3):142-4.

[10] Dujardin E, Mann S. Bio-inspired materials chemistry. Advanced Materials. 2002 Jun 5;14(11): 775-88.

[11] Kikuchi M, Itoh S, Ichinose S, Shinomiya K, Tanaka J. Self-organization mechanism in a bone-like hydroxyapatite/collagen nanocomposite synthesized in vitro and its biological reaction in vivo. Biomaterials. 2001 Jul 1;22(13):1705-11.

[12] Yamaguchi I. Preparation and mechanical properties of chitosan/ hydroxyapatite nanocomposites. Key Engineering Materials. 2001 Jul 1;192.

[13] Chen F, Wang ZC, Lin CJ. Preparation and characterization of nano-sized hydroxyapatite particles and hydroxyapatite/chitosan nano-composite for use in biomedical materials. Materials letters. 2002 Dec 1;57(4):858-61.

[14] Liao SS, Cui FZ, Zhang W, Feng QL. Hierarchically biomimetic bone scaffold materials: nano-HA/collagen/PLA composite. Journal of Biomedical Materials Research Part B: Applied Biomaterials: An Official Journal of The Society for Biomaterials, The Japanese Society for Biomaterials, and The Australian Society for Biomaterials and the Korean Society for Biomaterials. 2004 May 15;69(2):158-65.

[15] Deng X, Hao J, Wang C. Preparation and mechanical properties of nanocomposites of poly (D, L-lactide) with Ca-deficient hydroxyapatite nanocrystals. Biomaterials. 2001 Nov 1;22(21):2867-73.

[16] Jie, W., Yubao, L. and Weiqun, C., 2003. A study on nano-composite of hydroxyapatite and polyamide. Journal of materials science, 38(15), pp.3303-3306.

[17] Huang M, Feng J, Wang J, Zhang X, Li Y, Yan Y. Synthesis and characterization of nano-HA/PA66 composites. Journal of Materials

Science: Materials in Medicine. 2003 Jul;14(7):655-60.

[18] Hartgerink JD, Beniash E, Stupp SI. Self-assembly and mineralization of peptide-amphiphile nanofibers. Science. 2001 Nov 23;294(5547):1684-8.

[19] Marouf HA, Quayle AA, Sloan P. In vitro and in vivo studies with collagen/hydroxyapatite implants. International Journal of Oral & Maxillofacial Implants. 1990 Jun 1;5(2).

[20] Du C, Cui FZ, Zhu XD, de Groot KA. Three-dimensional nano-HAp/collagen matrix loading with osteogenic cells in organ culture. Journal of Biomedical Materials Research: An Official Journal of The Society for Biomaterials, The Japanese Society for Biomaterials, and The Australian Society for Biomaterials. 1999 Mar 15;44(4):407-15.

[21] Reneker DH, Chun I. Nanometre diameter fibres of polymer, produced by electrospinning. Nanotechnology. 1996 Sep 1;7(3):216.

[22] Kim GM, Wutzler A, Radusch HJ, Michler GH, Simon P, Sperling RA, Parak WJ. One-dimensional arrangement of gold nanoparticles by electrospinning. Chemistry of materials. 2005 Oct 4;17(20):4949-57.

[23] Dror Y, Salalha W, Khalfin RL, Cohen Y, Yarin AL, Zussman E. Carbon nanotubes embedded in oriented polymer nanofibers by electrospinning. Langmuir. 2003 Aug 19;19(17):7012-20.

[24] Kim GM, Michler GH, Pötschke P. Deformation processes of ultrahigh porous multiwalled carbon nanotubes/polycarbonate composite fibers prepared by electrospinning. Polymer. 2005 Aug 23;46(18):7346-51.

[25] Khalid P., Suman V. B., “Carbon nanotubes integrated nanocomposites membranes for purification of water”, Elsevier, Amsterdam, Netherlands, Aquananotechnology: Applications of Nanomaterials for Water Purification, 1st Edition, 2020, ISBN: 978-0-12-821141-0.

[26] Khalid P., Suman V. B., “Risk Management and Regulatory Aspects of Carbon Nanomaterials”, Elsevier, Amsterdam, Netherlands, Carbon Nanomaterials for Agri-food and Environmental Applications, 1st Edition, 2019, ISBN 978-01-28197-86-8.

[27] Khalid P., Suman V. B., “Quality Control and Risk Management of Carbon Nanomaterial”, INTECHOPEN LIMITED, London, UK, Perspective of Carbon Nanotubes, 2019, ISBN 978-1-78984-402-3.

[28] Khalid P., Suman V. B., “Current Toxicological Knowledge of Carbon Nanotubes”, Research India Publication, Advanced Engineering Research and Applications (Book Code: AERA), 2017, ISBN No. 978-93-84443-48-1.

[29] Khalid P., Hussain M. A., Rekha P. D., Suman V. B., Arun A. B., “Modifications of Carbon Nanotube for Bio-Applications and Toxicity Evaluation”, Bloomsbury Publication, India, Nano Biomaterials, 2012, ISBN 978-93-82563-37-2.

[30] Parwez K, Budihal SV. Optimisation of carbon nanotubes for advanced diagnosis and biomedical application. International Journal of Nanoparticles. 2019;11(3):217-38.

[31] Khalid P, Suman VB. Carbon nanotube-hydroxyapatite composite for bone tissue engineering and their interaction with mouse fibroblast L929 In Vitro. Journal of Bionanoscience. 2017 Jun 1;11(3):233-40.

[32] Koonprasert S, Channgam N. Global stability and sensitivity analysis of SEIQR worm virus propagation model with quarantined state in mobile

internet. Global Journal of Pure and Applied Mathematics. 2017;13(7):3833-50.

[33] Khalid, P., Hussain, M.A., Rekha, P.D. and Arun, A.B., 2015. Carbon nanotube-reinforced hydroxyapatite composite and their interaction with human osteoblast in vitro. Human & experimental toxicology, 34(5), pp.548-556.

[34] Parwez K, Budihal SV. Carbon nanotubes reinforced hydroxyapatite composite for biomedical application. Journal of Bionanoscience. 2014 Feb 1;8(1):61-5.

[35] Khalid P, Hussain MA, Rekha PD, Sanal C, Suraj S, Rajashekhar M, Suman VB, Arun AB. Interaction of Carbon Nanotubes Reinforced Hydroxyapatite Composite with Bacillus subtilis, *P. aeruginosa* and *C. albicans*. Indian Journal of Science and Technology. 2014 May 1;7(5):678.

[36] Khalid P, Hussain MA, Rekha PD, Suman V, Arun A. Modification of Carbon Nanotubes for bioapplications and toxicity evaluations. J. Environ. Nanotechnol. 2013 Jan 1;2(1):42-6.

[37] Khalid P, Hussain MA, Rekha PD, Arun AB. Synthesis and characterization of carbon nanotubes reinforced hydroxyapatite composite. Indian Journal of Science and Technology, 2013, 6(12), pp.5546-51. DOI: 10.1007/s12666-017-1150-6

Chapter 4

Advanced Carbon Materials for Sustainable and Emerging Applications

Aneeqa Bashir, Azka Mehvish and Maria Khalil

Abstract

Advanced carbon nanomaterials, which comprises fullerene, graphene, and carbon nanotubes (CNTs) are considered as backbone of engineering and scientific Innovation due to their versatile chemical, physical and electrical properties. Sustainable carbon materials are fabricated using different physical and chemical methods, respectively. Moreover, fabrication methods are used to achieve advanced carbon monoliths which are constituents with desirable properties. Keeping a view of desirable monoliths, diverse allotropes of advanced carbon nanomaterials are mostly employed in renewable energy resources, organic photovoltaic, and energy conservation technology, respectively. Carbon nanomaterials offer tremendous potential for enhancing biology and drug delivery because of biocompatibility. The proposed chapter grants a variety of fabrications methods for sustainable carbon materials as well as highlights the miscellaneous applications. Further, graphene, carbon nanotube (CNT) and fullerene are considered as fast and effective nanocarriers for delivering therapeutic molecules. As advanced carbon materials have controllable porous structure, high surface area, high conductivity, high temperature stability, excellent anti-corrosion property and compatibility in composite materials so they can be employed in energy storage as electrocatalysts, electro-conductive additives, intercalation hosts and ideal substrate for active materials. Meanwhile, the chapter sums up the required demands of advanced carbon materials for technological innovation and scientific applied research.

Keywords: Graphene, Drug Delivery, Porous Carbon, Physical Properties, Chemical or Physical Methods, Energy Storage, Catalysts

1. Introduction

Nowadays, Carbon science is immensely popular in several areas of material, nanoscience and, engineering [1] because of its unique and fundamental properties. Basically, the word carbon meaning Charcoaloriginates from Latin *carbo* [2]. It's among the most plentiful component on the earth. Since about 5000 BC, Carbon has been acknowledged and utilized for diverse purposes. From prehistoric times, carbon was primarily obtained from wood that has been used for fuel for millennium years. Graphite is the crystalline form of carbon that occurs naturally. The second allotrope of carbon is diamond and has been adopted as a gemstone. Diamond is extensively used in industry, due to its extraordinary optical properties [3].

In general, the 20th century is identified as plastics while 21st century will be named as 'century of graphene' due to its remarkable physical properties.

Graphite, diamond and amorphous carbon were only carbon allotropes known to exist in the early 1980s [4]. Where amorphous carbon does not have a crystalline structure and is used in various applications like ink, rubber filler, and paint. Carbon has a remarkable capability to bind with numerous other elements to make it essential to nearly all life. The current era is abounding with a variety of novel materials and undeniably the 20th-century capitulates a significantly more materials than previous centuries. Various materials have significant impact on society. Carbon has emerged as the most crucial substance for transforming 21st-century illumination for two reasons, carbon is immensely strong and very lightweight [5].

Discovery of fullerene developed the new aspect to the cognizance of carbon science, and it is the fourth allotrope of carbon at the nano level in which Spherical fullerene are buckyballs (C_{60}) is most common. In 1970, the existence of C_{60} was purposed by E. Osawa. Buckyball was discovered, in 1985 by R.F. Curl, and co-workers. In 1996, R.F. Curl and co-workers win Nobel Prize for this great discovery. Thereafter, in different natural environments, C_{60} was found in rock, on earth, and in space [6, 7]. In buckyballs, atoms are covalently bonded to three nearest atoms in closed-shell [3]. This serendipitous discovery gave rise to a new epoch for synthesized carbon allotropes. The next major breakthrough in this area was the carbon nanotubes synthesis mechanism. Carbon nanotubes were discovered by S. Iijima, in 1996 [8]. By using evaporated graphite, synthesis of semi uni-dimensional carbon tubes was reported forthwith the discovery of fullerenes [9]. Later on, by using the chemical vapor deposition fabrication technique, carbon nanotubes were developed [10]. One of the types of carbon nanotube is single-walled (SWNTs) in which consist of one atom thick layer while multi-walled (MWNTs) many layers of graphitic carbon in the wall [11]. Carbon nanotubes were discovered later but they give new dimensions towards the field of advanced carbon.

In 2004, A. Geim and his colleagues investigated and fabricated graphene, which is the final stage of succeeding progress of carbon-based material. In 2010, Andre Geim and his team won Nobel Prize for their innovative research on graphene. Graphene is considered as a primary constituent of various allotropes of carbon and knowing as a mother of carbon-based materials especially graphite [12]. Graphene has a honeycomb-like structure, that is made up of a single atom thick sp^2 hybridized carbon atom (**Figure 1**).

Diverse allotropes of carbon materials are mostly employed in renewable energy resources for the generation and storage of energy. Carbon nanomaterials

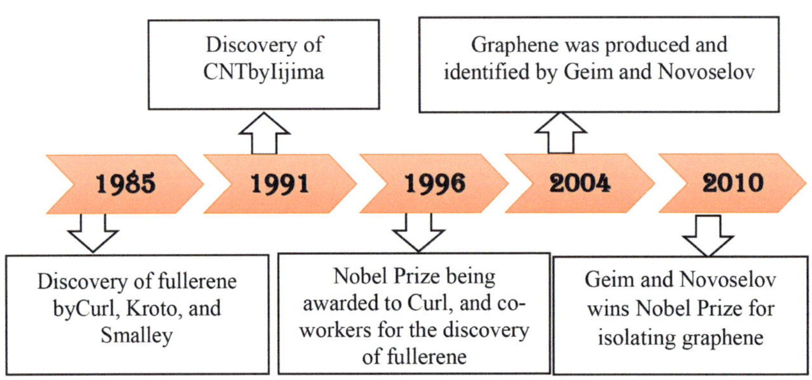

Figure 1.
Time line of advanced carbon materials.

offer tremendous potential for enhancing biology, medicine [13], electronics, energy technologies [14], and drug delivery. Scientists are currently working feverishly to achieve the inimitable electrical, mechanical, and thermal properties of their carbon materials for potential applications. Hence the interest in synthesis and applications of sustainable materials is emerging. Especially, for the sake of applications in future chemical and energy sector are becoming recognized. This chapter will introduce different fabrications methods for sustainable carbon materials as well as highlights their applications in essential energy and environmental-related field.

2. Fabrication of advanced carbon materials

With the innovation of industrial and commercial growth, the globe is confronted with a slew of crisis, including energy shortage and environmental issues, which have hampered the advancement of human civilizations. Researchers have employed a variety of methods to address these issues and have achieved certain results [15]. The synthesis for advanced carbon materials (Carbon nanotubes, Graphene and, fullerenes) has been developed tremendously. So, this material can fabricate by two methods physical methods and chemical methods. During the chemical production of nanoparticles, it is feasible to control particle size at a nanometer scale [16]. While the devoid of solvent and uniform nanoparticle dispersion is a fundamental aspect of this physical method over the chemical method [17].

3. Physical methods

3.1 Fabrication of carbon nanotubes via laser ablation

Smalley and his team established, method of laser vaporization for the preparationof fullerene as well as carbon nanotubes [18]. it was initially used for the synthesis of fullerene and then later it is used in the production of CNTs, particularly single-wall CNT [19]. Laser ablation is a technique that is used for the synthesis of carbon nanotubes. A furnace, graphite target, reactor tube and, laser beam source is used in this process (**Figure 2**).

A small graphite pellet is positioned at the mid of the quartz reactor tube inside the furnace at a specific temperature. Once the air from quartz has been evacuated, temperature of the furnace is raised to 1000°C- 1200°C. The quartz tube is saturated with inert gas. Scanning the surface of the pellet by beam of laser, graphite is vaporized whereas to preserve the smooth and homogenous surface. When gases

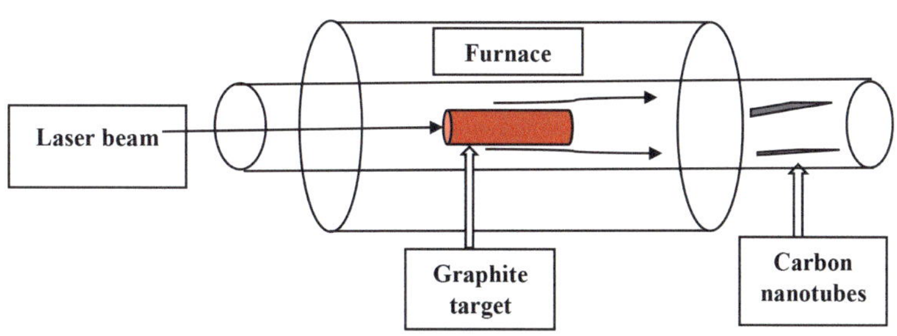

Figure 2.
Pictorial diagram of laser ablation method.

flow through chamber as a result, carbon constituents obtained via sublimation of graphite are displaced. Coalesce phenomena took place in gaseous state. After that, they deposit on the surface of the collector of cooled copper. The fundamental distinction between pulsed and continuous laser vaporization is that pulsed laser produced significantly higher power density. Many parameters influence the properties of CNTs synthesized by the pulsed laser deposition process (PLD), including laser properties, target material compositional and structural properties, chamber pressure, and chemical composition [5].

3.2 Fabrication of carbon nanotubes via electric arc discharge

By using arc discharge method, Zhai etl fabricated amazing form of carbon [11]. The AC/DC arc discharge technique may be used to prepared CNTs. The electric arc discharge synthesis method involves high voltage delivered between two graphene electrodes and high temperature. The synthesis process successfully accomplished by filling a chamber with inert gas, such as He/Ar gas. Potential difference is supplied among two water-cooled electrodes that are made up of graphite. Subsequently, gradually electrodes are brought closer together till the gap among the electrodes narrows reaches the point, where electric arc collapse happening. Afterward, sublimation of anode occurred, yielding the plasma at the region among the two electrodes, wherever temperature approaches 6000°C high enough to cause graphite sublimation (**Figure 3**).

The carbon atoms are evacuated from the solid during sublimation, the pressure is extremely high. Then these atoms migrate to the colder zone of the chamber, allowing a nanotube to deposit to accumulate on carbon. This type of nanotube is produced highly dependent on the composition of the anode. In comparison to other methods, this methodology leads to the growth of CNTs with few structural defects [5].

3.3 Synthesis of graphene via micromechanical exfoliation

Novoselov and Geim developed a very simple efficient method. This method involves, repeat the adhere and peel steps multiple times by using ordinary scotch tape, culminating a few mm -thick flakes of graphite of auni-layer thin specimen.

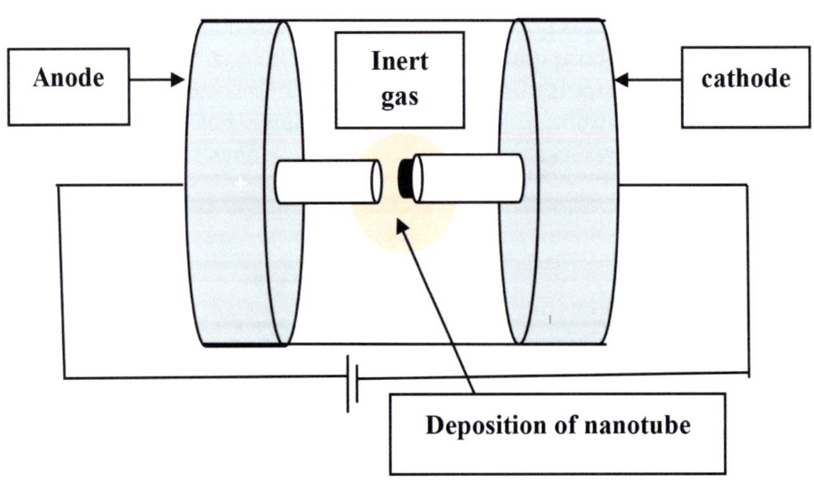

Figure 3.
Schematic diagram of Electric arc discharge.

By using manually mechanically cleavage graphite with scotch tape, the first graphene sheet was produced. Deifying the fact that no two-dimensional crystal can occur underneath the environmental conditions and, also displaying a variety of unique characteristics [20]. Mechanical exfoliation can produce high-quality graphene for characteristics research. But it not appears innovative for the huge areas. Mechanical exfoliation produced particles of graphene, with few microns of sideways dimension [21].

3.4 Synthesis of graphene via arc discharge method

Arc discharge techniquehas beenused to prepared sheets of graphene. For the first time, Rao et al. showed that this method may also be used to fabricate the sheets of graphene [22]. Here graphite is used as an electrode, they used comparatively high hydrogen pressure and produce graphene having 2–4 sheets in the innermost arc area. No catalytic agent has been employed in this process. Furthermore, doping of nitrogen and boron, sheets of graphene, can easily obtained by a nitrogen source (pyridine) and boron source (B_2H_6). Nevertheless, the graphene sheet's dimensions and shape can be optimized [23].

4. Chemical methods

4.1 Fabrication of carbon nanotubes via chemical vapor deposition (CVD)

CVD technique is utmost popular, uncomplicated as well as cost-effective for producing Carbon nanotubes, at relatively low temperature. In this approach, a gaseous carbon source is used. Transmit energy to the gaseous state of carbon molecules by using av ariety of energy sources such like a Heating coil or plasma source. As a source of carbon, CVD employing different hydrocarbons such as, carbon monoxide/methane. During this process, a substrate is coated with a metallic catalyst is heated around 700°C.When two different gases, one is actually carrier gases (Argon /Nitrogen/hydrogen) and, hydrocarbon transported via chamber, growth begins (**Figure 4**) [24].

When chamber is heated up using heating coil, then disintegrated gasses causing a reaction between reactive species from the gas and catalyst resulting in carbon deposition on a substrate. Through an elimination system, hydrogen gas as a residue will eliminate from the chamber [25]. The temperature range for the vapor deposition is 600°C-1200°C. The vapor of hydrocarbon will mix with a metallic catalytical agent till carbon and hydrogen are produced. While carbon will soluble in metallic substrate and evaporation of hydrogen will occur [23].

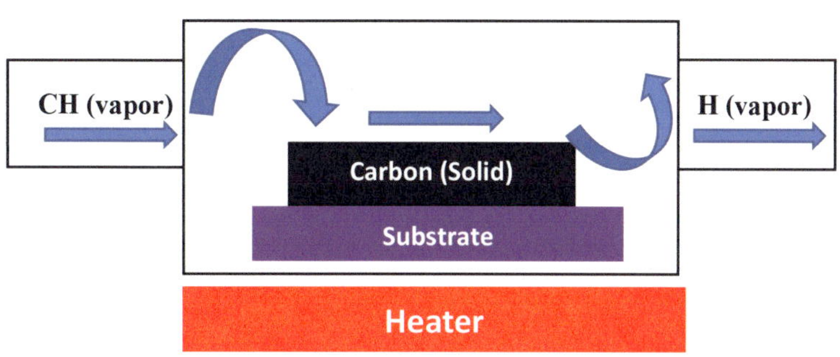

Figure 4.
Schematic diagram of CVD.

4.2 Fabrication of carbon nanotubes via hydrothermal method

The latest approach known as a hydrothermal method, use to synthesize the carbon nanotubes (CNTs) at low temperature, reducing the cost of CNTs production on a wide scale. Furthermore. Hydrothermal synthesis of materials has several advantages, environmentally Benefield, simple and quick. CNTs are made by utilizing the hydrothermal method, which uses ferrocene as carbon percussor and sulfur as a catalyst. Under constant stirring, sulfur and ferrocene have been mixed with NaOH solution containing ethanol and distilled water. Under the ultrasonic bath, the mixture has been sonicated at ambient temperature and constant stirring. The homogeneous solution is transferred to an autoclave reactor for hydrothermal treatment in an electric oven. Once the reaction is completed, the autoclave reactor has been cooled at room temperature. Black participate in the autoclave is filtered, and washed with ethanol, HCl, and then distilled water, until the pH 7 of washing solution is obtained [26].

4.3 Synthesis of graphene oxide (GO) via Hummer method

W.S. Hummers and R.E. Offeman developed this method in 1958 [27]. This method is appropriate for the synthesis of graphene oxide (GO) on wide-ranging. Hummer method emphasized three Phase responses [28]. Intercalation at a lower temperature around 5°C, moderate temperature approximately 35-40°C oxidation of graphite intercalation compounds, and hydrolysis due to higher temperature. For the preparation of GO, untainted graphite, concentrated sulfuric acid and, sodium nitrate is added to the flask. Kept that flask at temperature 5°C in an ice bath, under constant stirred for few minutes. After that, to avoid intense reaction at a certain spot, potassium permanganate has been slowly added to the flask. Next step, to remove the ice bath. Temperature of the reaction is gradually raised to a maximum temperature that is 35-40°C, and maintained for half an hour with continuous stirring. Demineralized water is then added to the mentioned suspension. As a result, hydration of heat, producing a large exotherm of approximately 98°C. After that bath process is maintained at maximum temperature. Later on, deionized water and hydrogen peroxide have

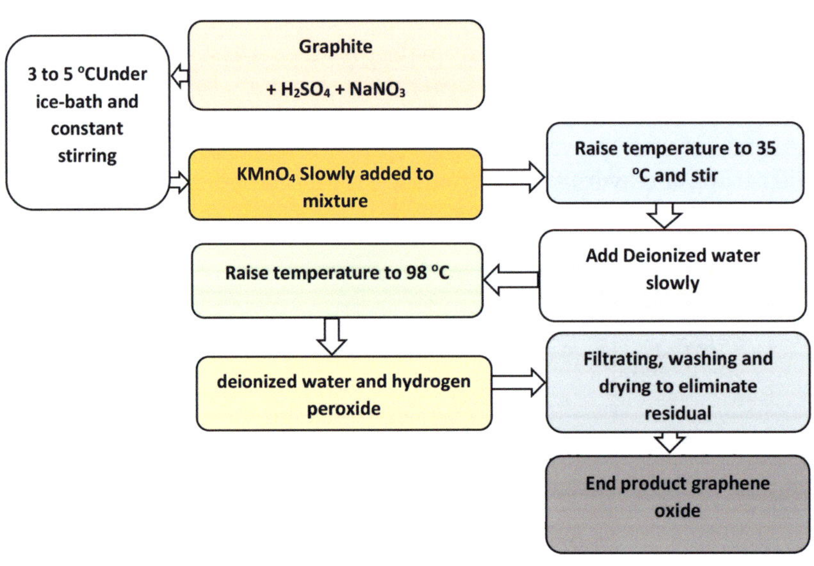

Figure 5.
Schematic flow chart of GO synthesis by Hummer's Method.

been added to finish the reaction. Vacuum filtration extracts the finished product from the solution, which is in yellowish/brown color. The GO granules are then rinsed five times in dilute hydrochloric acid. and warm deionized water to eliminate any residual Mn ions or acid [29]. The graphene oxide (GO) has been synthesized via filtering and drying under the vacuum at room temperature (**Figure 5**).

4.4 Synthesis of graphene oxide (GO) via modified Hummer method

Modified Hummer method is used to synthesized GO. In this procedure, graphite is added to concentrated sulfuric acid and phosphoric acid (relatively 9:1 ratio of volume) [30]. Stirred for 10–20 minutes in bath ice. A double amount of potassium permanganate has been gradually added into the mixture under stirring. Keeping temperature of solution around 3-5°C solution. The mixture has been reacted for 2–4 hours in a cold bath. Once again mixture has been stirred, at a 38–40°C water bath. Consequently, for one hour, the temperature of the mentioned mixture is maintained at 98°C [31]. Deionized water is continually added. Along with hydrogen peroxide also has been added dropwise in the said suspension. After filtration, mixture is washed with deionized water and Hydrochloric acid multiple times. At last, the end product is dried to get final outcome of GO (**Figure 6**).

4.5 Fabrication of reduced graphene oxide (rGO) via Tour method

Tour's group developed an improved version of Hummer's method in 2010 at Rice University. In the mixture of H_2SO_4/ H_3PO_4 (9:1), they replaced the sodium nitrate with orthophosphoric acid and increased the quantity of $KMnO_4$. There is no generation of hazardous gas such as NO_2, N_2O_4, or $ClO_2$are regarded as the benefit of this method, and temperature is easily controllable in Tour's method. According to the author, the existence of orthophosphoric acid leads more intact graphitic basal plane [32]. Tour's method is most efficient for GO as, it is inexpensive, non-toxic and also, environmentally benign [33].

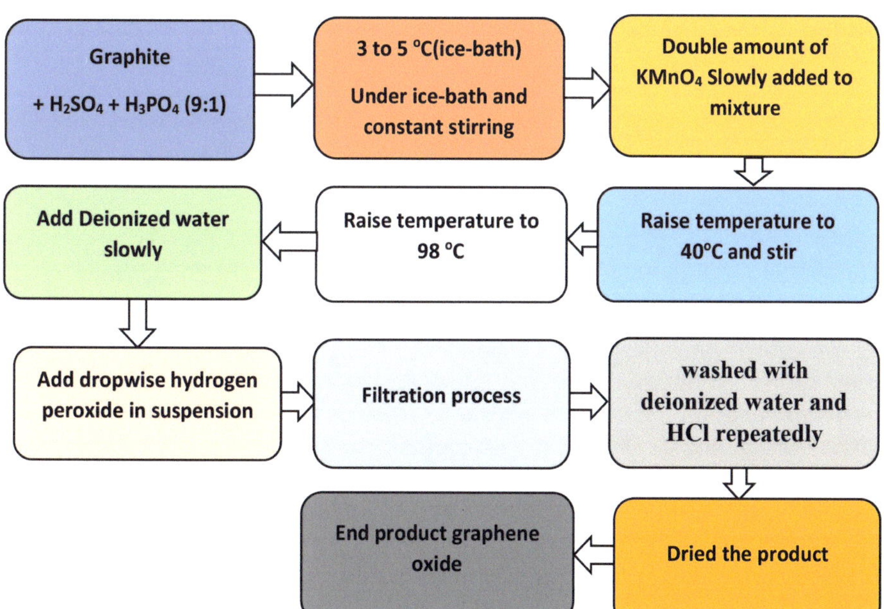

Figure 6.
Schematic flow chart of GO synthesis by Modified Hummer's Method.

5. Applications of advanced carbon materials

Advanced carbon nanomaterials are receiving a lot of attention of scientific research in last few years owing to their unique mechanical, thermal, chemical, optical and electrical properties. Advanced carbon materials, comprising of graphene, fullerene, carbon fibers, activated carbon and carbon nanotubes are considered as backbone of material science and technological innovation [34]. These advanced carbon materials also find applications in electronics, organic photovoltaic, energy conservation technology and drug delivery as illustrated in **Figure** 7 [35].

5.1 Drug delivery

The advancement of drug delivery systems is appealing as it permits to improve the therapeutic properties of already existing drugs. The main purpose of developing an improved drug delivery system (DDS) is to transport therapeutic agents to the diseased area in a controlled way with few side-effects on other healthy tissues [36]. In the last few years, the advancement in nanotechnology has avoided some problems in this area by inserting nanosacled drug carriers with useful applications in drug delivery system (DDS). Among different nano-sized drug carrier, advanced carbon materials have been extensively discussed for delivering therapeutic molecules, owing to their advantageous physical as well as chemical characteristics [37].

Recently, graphene has been explored as novel and inexpensive DDS with the possibility of being employed for systemic, targeting and local DDS [38]. Graphene is considered as one of the best nanocarrier for drug delivery because of its many reasons. Firstly, it can load more drugs than other nanocarriers owing to larger surface provided by arrangement of atoms in two dimensions. Secondly, graphene and GO are suitable for different delivery environments due to their high chemical and mechanical stability. Thirdly, toxicity and side effects of graphene and GO can be reduced by simple functionalization process such as coating with biocompatible material [39]. As graphene has an excellent ability to deliver protein, peptide and nucleic acid into the cell by crossing the cell membrane so it has been explored for delivering various therapeutics drugs, including antibodies, genes, antibiotics, anti

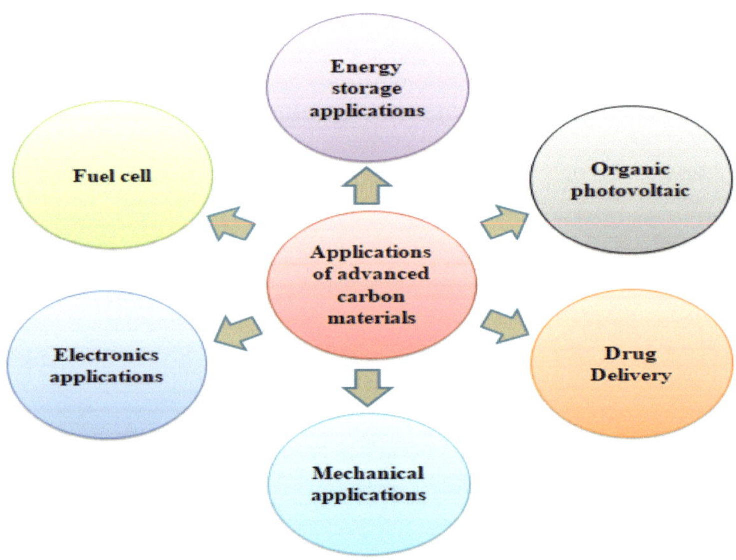

Figure 7.
Advanced carbon material's applications.

cancer medicine, RNA, DNA and insoluble drugs. Delivery of multiple drugs at once by graphene based nanomaterials has also been investigated for chemotherapeutic efficacy [40].

In addition to graphene, carbon nanotube (CNT) has been investigated as fast and effective nanocarrier for delivering therapeutic molecules. It has become an attractive task for many research groups for its excellent drug delivery applications owing to its unique biological, physical as well as chemical properties [41]. CNT also has ability to attach desired functional groups on its outer layer due to its hollow monolithic structure, which makes the CNT a promising nanocarrier for drug delivery. Furthermore, functionalized CNTs can easily penetrate into the cells so they have potential to work as vehicles for drug delivery of small therapeutic molecules [42].

In addition, fullerene is also considered a promising material for drug delivery in the body as catalyst and as a lubricant. It is mostly used to deliver the drugs for cancer therapy, drugs needed to enter the brain and -tumor drugs. It acts as hollow cages to capture other molecules. That is how it delivers drug molecules to the site of action inside the body [43]. It can also transport multiple drugs to different tumors by combining monoclonal antibody. Fullerene is efficient for delivering the drugs owing to its capability to pass blood through brain barrier, deliver directly to the targeted cells, not dissolve before reaching the targeted region and carry the load of multiple drugs [44].

5.2 Energy storage applications

In recent years the demand for highly efficient energy storage devices has increased significantly and a lot of work has been done to develop advanced electrode materials. In this respect, advanced carbon materials have been widely investigated for sustainable clean-energy storage systems owing to their excellent physicochemical and thermo-mechanical characteristics [45]. As advanced carbon materials have controllable porous structure, high surface area, high conductivity, high temperature stability, excellent anti-corrosion property and compatibility in composite materials so they can be used in energy storage devices as electrocatalysts, electro-conductive additives, intercalation hosts and ideal substrate for active materials [46].

Within family of advanced carbon materials, graphene is attracting a lot of scientific attention for its remarkable characteristics including significant surface area, mechanical stability and excellent electrical conductivity, making it suitable to use as electrodes in energy storage systems [47]. They can used to increase the efficiency of currently used energy storing devices for instance super-capacitor and Lithium ion batteries as well as make next generation devices such as Sodium ion batteries, Lithium sulfur batteries and Lithium O_2 batteries, more practical. In supercapacitor, in order to form an effective electric double layer coating, graphene is used in the carbon coatings. These supercapacitors are then utilized to store large amount of energy. Graphene is also used as electrode material in supercapacitor and pseudocapacitors [48].

Carbon nanoube is also another promising material to be used in different energy storage applications. It is used as electrodes in supercapacitor and lithium ion batteries owing to its extraordinary tensile strength, electrical conductivity as well as ultrahigh surface region. Energy storage devices employ CNTs as additives in order to increase electrical conductivity of cathode and also as anode components [49, 50].

Apart from graphene and CNTs, activated carbon also finds its potential as electrode material in energy storage applications due to its wide surface region and excellent electronic conductivity. These properties can be increased by adding other materials such as polymer and other carbon based materials [47].

Fullerene is also used in advanced energy storing devices as electrodes because of their being lightweight, controllable electrochemical performance, flexibility and excellent electrical conductivity [51].

5.3 Fuel cell applications

Electrochemical fuel cell technology is a source of continuous supply of energy and is important for providing sustainable energy conversion system. Fuel cell has become a unique device for energy storage applications varying from mobile phones to power plants because of their excellent efficiency, high performance and less pollutant's emission [52]. For development of fuel cells, one of the crucial components of fuel cell is catalyst support. Catalyst support determines the level of catalyst dissemination and increases the catalyst reactivity. The main characteristics of catalyst support include large surface area, excellent thermal and electrical conductor, stable in different types of working media and good absorbent to allow reactant movement. Catalyst support is necessary to improve fuel cell's efficiency [53]. All these properties are possessed by advanced carbon materials. In fuel cells, advanced carbon materials are used as catalyst as well as catalyst support in order to enhance cell performance. As compare to other materials, advanced carbon materials have several advantages, such as good stability in alkaline and acidic media, electric conductivity and wide surface region. Moreover, these advanced nanomaterials are also used in membranes electrode assemblies (MEA) in fuel cell applications [52, 54].

Among these advanced carbon materials, graphene is favorable for energy conversion systems owing to its certain characteristics including high tensile strength, ultrahigh surface area, chemically stable as well as electronic conductivity, making it a potential candidate for fuel cell applications [55]. Due to all these extraordinary properties, graphene is mostly employed as electro-conductive electrode, catalyst support, bipolar plates and additives in fuel cell technology as shown in **Figure 8**.

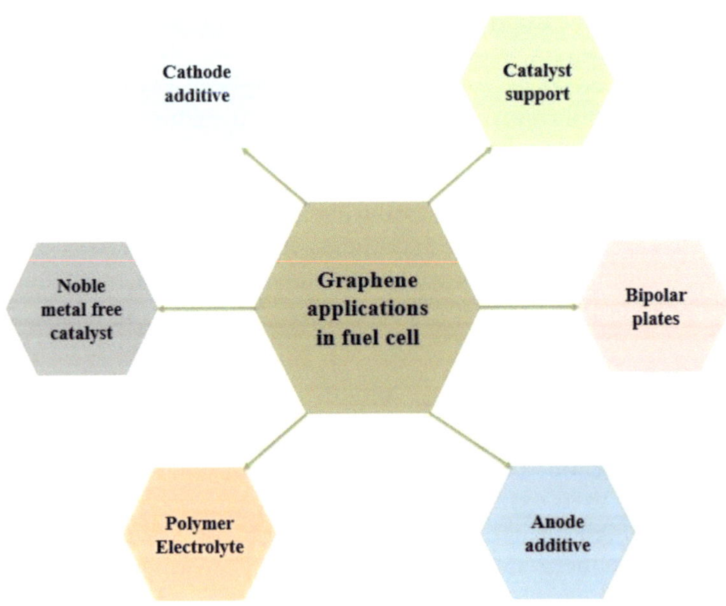

Figure 8.
Graphene's applications in fuel cell technology.

As an additive, graphene is used in electrodes either individually or mix with other carbon based materials [56]. They are incorporated in electrodes (anode and cathode) in order to enhance reactions in fuel cell. So in this way graphene offers its potential as catalyst to enhance fuel cell performance [57].

Besides graphene, carbon nanotubes are also favorable material for fuel cell applications. Properties of multi walled carbon nanotubes (MWCNTs) can be modified via functionalization process in order to produce good combination and improve performance to be used as catalyst support. Furthermore, they can be incorporated in the mixture of platinum/carbon catalyst at the anode to increase the efficiency of the catalytic reactions in the fuel cell [54]. Nitrogen-doped carbon nanotubes have been employed in fuel cells in order to decrease the oxygen level at the cathode. Carbon nanotubes in fuel cells are also used to reduce the demand of metals that are utilized as catalyst as well as enhance the fuel cell's efficiency.

As a catalyst support, carbon nanotubes have efficiently improved the catalyst performance and usage. The fuel cell employing carbon nanotubes as a catalyst support possess larger current density and high performance as well as catalytic activity [58].

In fuel cell technology, fullerene also finds its potential applications as electrocatalyst support at the anode for fuel oxidation reaction, for oxygen reduction reactions occurring at cathode and as proton conducting membrane [59].

5.4 Organic photovoltaic applications

Recently, organic photovoltaic (OPVs) have gain the interest of many research groups because of their usage as conformal, flexible, lightweight and inexpensive power supplies for different commercial applications. OPVs are most effective transformative solar technology made of earth abundant and nontoxic materials on larger scale. Their performance for commercial applications can be improved despite their high capability in large scale development [60]. In order to meet these performance goals, advanced carbon based materials are extensively used as transparent flexible electrodes, hole transporters and electron acceptor in OPV devices. Advanced carbon-based photovoltaic devices have attracted much attention for both scientific research and commercial applications. Advanced carbon nanomaterials namely fullerenes, graphene, activated carbon and carbon nanotubes possess attractive properties to be employed as active materials for the development of OPV devices [61].

In OPV devices, organic solar cell is most extensively used device for potential applications. The aim of organic solar cell is to provide effective and low cost energy production photovoltaic device as compared to already present solar technologies. Generally, organic solar cell is a very simple device made up of small organic molecules, polymers, or mixture of both materials with or without other nanomaterials added into whole device [62]. These cells are not only flexible but also their fabrication cost is just one third of the price of silicon cell. They can be inserted in infrastructures including walls, windows, car windshields and many others [60].

Although graphene has single atom thick sheet structure but due to high hole transport movement and larger surface region, graphene finds its applications in organic solar cells [30]. In organic solar cells, graphene thin films are extensively utilized as window electrodes. There are many advantages of using graphene in these cells. Firstly, it generates a window for inducing photon energy in a wide range of wavelengths (from UV to far infrared) inside solar cells. Secondly, it builds a flexible transparent device with robust architecture. Thirdly, it possesses high charge transfer mobility at the interface of electrochemical cells and allows for more heat dissipation [63].

In addition to graphene, Carbon nanotubes (CNTs) have unique characteristics, such as excellent electrical conductivity, that make them ideal for combining with conducting polymers to create composites being used in organic solar cell. These cells have been made by incorporating carbon nanotubes into conjugated polymers. In principle, carbon nanotubes and their composites can replace all parts of a solar cell, including the light sensitive component, carrier selective contacts, passivation layers, and transparent conducting films [64]. Over the last decade, carbon nanotube has been employed in organic solar cell as additives, carrier transporters, transparent electrodes and light absorbers. Single walled carbon nanotube has been widely utilized as an electron donor with fullerene-derivates as an acceptor in organic solar cell using carbon nanotubes in photoactive layer to produce a type II heterojunction [65].

With the discovery of photo-induced electrons transfer from conducting polymers to fullerene, fullerene has been the most favored electron accepting and transferring material for organic solar cell. Importantly, to improve the organic solar cell's efficiency, the thickness of photoactive layer can be decreased by utilizing single-crystal fullerene [66].

5.5 Electronics applications

Advanced carbon materials are well recognized for their usage in electronics because of their distinctive properties. Among advanced carbon materials, graphene is the outstanding material with its attractive characteristics which make it preferable choice for advanced electronics applications. Graphene finds its potential to bring revolution in electronics due to its remarkable strength, thermal and electrical conductivity and electron mobility [67]. In electronics graphene is mostly used as transistor, interconnect, sensor and thermal management. It's one of the most efficient electrical conductors on the Earth.

In electronics and integrated circuits, the characteristics of graphene have made it a viable alternative to silicon and the basis for developing superconductors which permit to generate higher voltage lines that will effectively transfer energy to our houses [68]. Microchips and transistors, both fundamental components in almost all electrical devices, can be made from graphene. Graphene is used to build circuitry of computer resulting in the development of significantly faster computers with low power consumption than existing silicon. It is used to build circuitry Touch screens for cell phones and tablets can be improved by using graphene as a coating [69].

Besides graphene, carbon nanotubes for electronic applications are also interesting topic for research and printable carbon nanotube inks are making their way to the market. Carbon nanotubes are utilized in the manufacture of transistors and are used as conductive layers in the rapidly expanding touch screen sector. In some applications, carbon nanotubes are thought to be a good substitute for Indium tin oxide (ITO) transparent conductors [70]. For instance, carbon nanotubes, when fabricated as transparent conductive films (TCF), might be utilized as transparent, highly conductive and inexpensive alternative in flexible displays and touch panels. Significant development has been achieved on carbon nanotubes field effect transistor (FET). Carbon nanotube-based field effect transistors (FETs) exhibit operating characteristics that are almost comparable to those of silicon-based components [71].

Single-wall carbon nanotube (SWCNTs) is promising candidate for the channel material of flexible thin film transistor (TFT). Thermal management of electrical circuits is achieved using large structure of carbon nanotubes. Random networks of SWCNTs have also been used to fabricate integrated logical circuits. Carbon nanotubes also act as a rectifying diode [72]. In molecular electronics carbon nanotube is good candidate for connection because of their structure, electronic conductivity as well as capability to be precisely generated. It is also recently revealed that SWNTs, when employed as interconnects on semiconductor devices, they can transmit electrical signals at high speed [73]. Fullerene is also used in electronics in a variety of ways, including as a diode, transistor, and photocell. Moreover, fullerene also behaves as device interconnects [74].

5.6 Lightweight mechanical applications

One of the most intriguing properties of graphene is that it is the lightest and strongest substance known. It is lighter than aluminum and more elastic than rubber. The reason that graphene stands out as a reinforce ingredient in composite material is its outstanding intrinsic mechanical characteristics, such as stiffness, strength, and toughness [75]. Currently, graphene is expected to be used (possibly incorporated with plastic) to make a material that would substitute steel in airplane structures, improving its performance, range, and weight. Because of its electric conductivity, it is utilized to cover airplane outer surface to protect it from electrical damage. This covering is also utilized to determine strain rate, alerting the pilots to any variations in the stress levels on the airplane's wings. The advancement in high strength demanding potential applications just like body armor for armed persons is possible with the help of useful properties of graphene [76]. Graphene is a promising material for the development of speakers as well as electrostatic audio microphones because of its lightweight property, which in turns show fairly good frequency response. For applications considering strength and weight as limiting factors such as in aerospace industry, graphene can be added into a variety of composites. It is incorporated in different materials in order to make existing materials more stronger and lightweight. So a composite material that is more lighter and stronger than steel, gives the necessary strength required for the aviation industry and save more money on fuel consumption. This is the main reason of incorporating graphene into these materials [77, 78].

Carbon nanotubes also possess extraordinary mechanical properties. The Young's modulus of carbon nanotubes is about 1TPa and their tensile strength is about 11–63 GPa. Due to these unique properties, carbon nanotubes have gained much attention for many mechanical applications such as rotational actuators, nanometer cargoes, high frequency oscillators and nanometer tweezers. Extraordinary mechanical properties of carbon nanotubes make it preferable material for load-bearing reinforcement in composite materials as well as also for structural applications [79]. Moreover, low specific weight and high young modulus of SWCNT enable them to employ as ultimate mechanical resonators for analyzing mechanical motion in quantum field. Tip used in scanning probe microscopy can be made of carbon nanotubes due to its high elasticity property that avoids the mechanical destruction of tip when in contact with the substrate. Carbon nanotubes are also widely used in various sensor applications such as in mechanical sensors [80].

6. Conclusions and future directions

Advanced carbon nanomaterials are receiving a lot of attention of scientific research in last few years owing to their unique mechanical, thermal, chemical, optical and electrical properties. Advanced carbon nanomaterials, comprising of graphene, fullerene, carbon fibers, activated carbon and carbon nanotubes are considered as backbone of material science and technological innovation. These nanomaterials are fabricated by using different physical and chemical methods to get high materials with excellent characteristics. Advanced carbon materials also find applications in electronics, organic photovoltaic, energy conservation technology and drug delivery etc. In future these advanced materials can be used to develop several materials with different applications. A lot of research is taking place for producing these materials on industrial level. These advanced materials are the future of sustainable energy production and storage devices owing to its capability to store energy on large scale. Fuel cells also in the near future are thought to replace battery based energy systems. Graphene sheets may have the potential to be a game-changing use in microelectronics. The demand for advanced carbon materials will be further growing for technological innovation.

Author details

Aneeqa Bashir*, Azka Mehvish and Maria Khalil
Department of Physics, University of the Punjab, Lahore, Pakistan

*Address all correspondence to: aneeqa.physics@pu.edu.pk

IntechOpen

© 2021 The Author(s). Licensee IntechOpen. This chapter is distributed under the terms of the Creative Commons Attribution License (http://creativecommons.org/licenses/by/3.0), which permits unrestricted use, distribution, and reproduction in any medium, provided the original work is properly cited. (cc) BY

References

[1] Nasir, S., Hussein, M. Z., Zainal, Z., & Yusof, N. A. (2018). Carbon-based nanomaterials/allotropes: A glimpse of their synthesis, properties and some applications. Materials, *11*(2), 295.

[2] H.W. Fowler, F.G. Fowler (eds.), The Concise Oxford Dictionary of Current English (Clarendon Press, Oxford, 1964)

[3] Ren, Z., Lan, Y., & Wang, Y. (2012). *Aligned carbon nanotubes: physics, concepts, fabrication and devices*. Springer Science & Business Media.

[4] Hirsch, A. (2010). The era of carbon allotropes. Nature materials, *9*(11), 868-871.

[5] HAick, H. (2013). Nanotechnology and Nanosensors Introduction to Nanotechnology. Technion Israel Institute of technology ebook version, 12.

[6] P.R. Buseck, S.J. Tsipursky, R. Hettich, Fullerenes from the geological environment. Science257(5067), 215-217 (1992)

[7] J. Cami, J.Bernard-Salas, E. Peeters, S.E.Malek,Detection ofC60 andC70 in a young planetary nebula. Science 329(5996), 1180-1182 (2010)

[8] S. Iijima, and T. Ichihashi, Single-shell carbon nanotubes of 1-nm diameter, Nature, 1993. 363(6430): p. 603-605.3.

[9] Hirsch, A. &Vostrowsky, O. Top. Curr. Chem. 245, 193-237 (2005).

[10] Iijima, S. Nature 354, 56-58 (1991).

[11] Zhai, Y., Dou, Y., Zhao, D., Fulvio, P. F., & Mayes, R. T. (2011). S. Dai. Adv. Mater., 23, 4828.

[12] A.K. Geim, and K.S. Novoselov. The rise of graphene. Nature Mater. 6,183-191 (2007).

[13] L. Zhang and T.J. Webster, Nanotechnology and nanomaterials: Promises for improved tissue regeneration, NanoToday, 4(1), pp. 66-80, 2009.

[14] H. Wang, X. Yuan, Y. Wu, H. Huang, X. Peng, G. Zeng, H. Zhong,J. Liang, M.-M. Ren, Graphene-based materials: Fabrication, characterizationand application for the decontamination of wastewater andwastegas and the hydrogen storage/generation, Advances in Colloidand Interface Science (2013).

[15] Pan, Y., Liu, X., Zhang, W., Liu, Z., Zeng, G., Shao, B.,... & Chen, M. (2020). Advances in photocatalysis based on fullerene C60 and its derivatives: Properties, mechanism, synthesis, and applications. Applied Catalysis B: Environmental, *265*, 118579.

[16] Hyeon T (2003) Chemical synthesis of magnetic nanoparticles. Chem Commun. 927-934. Ingle A, Gade A, Pierrat S et al (2008) Mycosynthesis of silver nanoparticles using the fungus *Fusarium acuminatum* and its activity against some human pathogenic bacteria. CurrNanosci 4:141-144. doi:10.2174/157341308784340804

[17] Mageswari, A., Srinivasan, R., Subramanian, P., Ramesh, N., &Gothandam, K. M. (2016). Nanomaterials: classification, biological synthesis and characterization. In *Nanoscience in Food and Agriculture 3* (pp. 31-71). Springer, Cham.

[18] Guo, T., Diener, M.D., Chai, Y., Alford, M.J., Haufler, R.E., Mcclure, S.M., Ohno, T., Weaver, J.H., Scuseria, G.E., Smalley, R.E.: Science 257, 1661-1664 (1992)

[19] Thess, A., Lee, R., Nikolaev, P., Dai, H.J., Petit, P., Robert, J., Xu, C., Lee, Y.H., Kim, S.G., Rinzler, A.G., Colbert, D.T., Scuseria, G.E., Tomanek, D.,

Fischer, J.E., Smalley, R.E.: Science 273, 483-487 (1996)

[20] Tiwari, A., & Shukla, S. K. (Eds.). (2014). *Advanced carbon materials and technology*. John Wiley & Sons.

[21] Wang, Z. (2017, September). Progress on preparation of graphene and its application. In *IOP Conference Series: Materials Science and Engineering* (Vol. 242, No. 1, p. 012032). IOP Publishing.

[22] C.N.R. Rao, A.K. Sood, K.S. Subrahmanyam, A. Govindaraj, Epitaxial graphite: 2D electron gas properties and a route toward graphenebased, Angew. Chem. Int. Ed. 48, 7752-7777, 2009

[23] Venkataraman A, Amadi E V, Chen Y and Papadopoulos C 2019 *Nanoscale Res.* Lett. 14 220

[24] Khan, W., Sharma, R., & Saini, P. (2016). Carbon nanotube-based polymer composites: synthesis, properties and applications. *Carbon Nanotubes-Current Progress of their Polymer Composites*.

[25] Kumar M and Ando Y 2010 J. Nanosci. Nanotechnol. 10 3739

[26] Razali, M. H., Ahmad, A., Azaman, M. A., & Amin, K. A. M. (2016). Physicochemical Properties of Carbon Nanotubes (CNT's) Synthesized at Low Temperature using Simple Hydrothermal Method. International Journal of Applied Chemistry, *12*(3), 273-280.

[27] Hummers Jr, W. S., & Offeman, R. E. (1958). Preparation of graphitic oxide. Journal of the american chemical society, *80*(6), 1339-1339.

[28] Fu L, Liu Hongbo, Zou Yanhong, Li Bo. Technology research on oxidative degree of graphite oxide prepared by Hummers method (in Chinese). CARBON 2005;124(4):10-14.

[29] Guerrero-Contreras, J., & Caballero-Briones, F. (2015). Graphene oxide powders with different oxidation degree, prepared by synthesis variations of the Hummers method. Materials Chemistry and Physics, *153*, 209-220.

[30] Zaaba, N. I., Foo, K. L., Hashim, U., Tan, S. J., Liu, W. W., &Voon, C. H. (2017). Synthesis of graphene oxide using modified hummers method: solvent influence. Procedia engineering, *184*, 469-477.

[31] Song, J., Wang, X., & Chang, C. T. (2014). Preparation and characterization of graphene oxide. Journal of Nanomaterials, *2014*.

[32] Thakran, M., & Kumar, S. R. B. Recent Advances in Synthetic Methods to Prepare Graphene Oxide.

[33] Habte, A. T., &Ayele, D. W. (2019). Synthesis and characterization of reduced graphene oxide (rGO) started from graphene oxide (GO) using the tour method with different parameters. *Advances in Materials Science and Engineering*, *2019*.

[34] Ikram, M., Raza, A., Shahzad, K., Haider, A., Haider, J., Durrani, A. K.,... & Ikram, M. (2021). Advanced Carbon Materials: Base of 21st Century Scientific Innovations in Chemical, Polymer, Sensing and Energy Engineering.

[35] Tiwari, A., & Shukla, S. K. (Eds.). (2014). Advanced carbon materials and technology. John Wiley & Sons.

[36] Safari, J., & Zarnegar, Z. (2014). Advanced drug delivery systems: Nanotechnology of health design A review. Journal of Saudi Chemical Society, 18(2), 85-99.

[37] Tiwari, H., Karki, N., Pal, M., Basak, S., Verma, R. K., Bal, R.,... & Sahoo, N. G. (2019). Functionalized graphene oxide as a nanocarrier for dual

drug delivery applications: The synergistic effect of quercetin and gefitinib against ovarian cancer cells. Colloids and Surfaces B: Biointerfaces, 178, 452-459.

[38] Goenka, S., Sant, V., & Sant, S. (2014). Graphene-based nanomaterials for drug delivery and tissue engineering. Journal of Controlled Release, 173, 75-88.

[39] Wang, Z., Colombi Ciacchi, L., & Wei, G. (2017). Recent advances in the synthesis of graphene-based nanomaterials for controlled drug delivery. Applied Sciences, 7(11), 1175.

[40] Liu, J., Cui, L., & Losic, D. (2013). Graphene and graphene oxide as new nanocarriers for drug delivery applications. Acta biomaterialia, 9(12), 9243-9257.

[41] Kaur, J., Gill, G. S., & Jeet, K. (2019). Applications of carbon nanotubes in drug delivery: A comprehensive review. Characterization and biology of nanomaterials for drug delivery, 113-135.

[42] Bianco, A., Kostarelos, K., & Prato, M. (2005). Applications of carbon nanotubes in drug delivery. Current opinion in chemical biology, 9(6), 674-679.

[43] Dellinger, A., Zhou, Z., Connor, J., Madhankumar, A. B., Pamujula, S., Sayes, C. M., & Kepley, C. L. (2013). Application of fullerenes in nanomedicine: an update. Nanomedicine, 8(7), 1191-1208.

[44] Melanko, J. B., Pearce, M. E., & Salem, A. K. (2009). Nanotubes, nanorods, nanofibers, and fullerenes for nanoscale drug delivery. In Nanotechnology in drug delivery (pp. 105-127). Springer, New York, NY.

[45] Wei, Q. (2018). Advanced Carbon-Based Materials for Energy Storage and Conversion Applications (Doctoral dissertation, Université du Québec, Institut national de la recherche scientifique).

[46] Wu, M., Liao, J., Yu, L., Lv, R., Li, P., Sun, W.,... & Ma, J. (2020). 2020 roadmap on carbon materials for energy storage and conversion. Chemistry–An Asian Journal, 15(7), 995-1013.

[47] Iqbal, S., Khatoon, H., Pandit, A. H., & Ahmad, S. (2019). Recent development of carbon based materials for energy storage devices. Materials Science for Energy Technologies, 2(3), 417-428.

[48] Olabi, A. G., Abdelkareem, M. A., Wilberforce, T., & Sayed, E. T. (2021). Application of graphene in energy storage device–A review. Renewable and Sustainable Energy Reviews, 135, 110026.

[49] Amin, R., Kumar, P. R., & Belharouak, I. (2020). Carbon Nanotubes: Applications to Energy Storage Devices. In Carbon Nanotubes-Redefining the World of Electronics. IntechOpen.

[50] Lee, Y. H., An, K. H., Lim, S. C., Kim, W. S., Jeong, H. J., Doh, C. H., & Moon, S. I. (2002). Applications of Carbon Nanotubes to Energy Storage Devices (NDFCT 396). New Diamond and Frontier Carbon Technology, 12(4), 209-228.

[51] Ni, J., & Li, Y. (2016). Carbon nanomaterials in different dimensions for electrochemical energy storage. Advanced Energy Materials, 6(17), 1600278.

[52] Luo, C., Xie, H., Wang, Q., Luo, G., & Liu, C. (2015). A review of the application and performance of carbon nanotubes in fuel cells. Journal of Nanomaterials, 2015.

[53] Samad, S., Loh, K. S., Wong, W. Y., Lee, T. K., Sunarso, J., Chong, S. T., &

Daud, W. R. W. (2018). Carbon and non-carbon support materials for platinum-based catalysts in fuel cells. international journal of hydrogen energy, 43(16), 7823-7854.

[54] Bruno, M. M., & Viva, F. A. (2014). Carbon materials for fuel cells. In Direct Alcohol Fuel Cells (pp. 231-270). Springer, Dordrecht. [22] C.N.R. Rao, A.K. Sood, K.S. Subrahmanyam, A. Govindaraj, Epitaxial graphite: 2D electron gas properties and a route toward graphenebased, Angew. Chem. Int. Ed. 48, 7752-7777, 2009

[55] Hur, S. H., & Park, J. N. (2013). Graphene and its application in fuel cell catalysis: a review. Asia-Pacific Journal of Chemical Engineering, 8(2), 218-233.

[56] Iqbal, M. Z., Rehman, A. U., & Siddique, S. (2019). Prospects and challenges of graphene based fuel cells. Journal of Energy Chemistry, 39, 217-234.

[57] Su, H., & Hu, Y. H. (2021). Recent advances in graphene-based materials for fuel cell applications. Energy Science & Engineering, 9(7), 958-983.

[58] Akbari, E., & Buntat, Z. (2017). Benefits of using carbon nanotubes in fuel cells: a review. International Journal of Energy Research, 41(1), 92-102.

[59] Coro, J., Suarez, M., Silva, L. S., Eguiluz, K. I., & Salazar-Banda, G. R. (2016). Fullerene applications in fuel cells: A review. International Journal of Hydrogen Energy, 41(40), 17944-17959.

[60] Kim, T. H., Yang, S. J., & Park, C. R. (2011). Carbon nanomaterials in organic photovoltaic cells. Carbon letters, 12(4), 194-206.

[61] Zhu, H., Wei, J., Wang, K., & Wu, D. (2009). Applications of carbon materials in photovoltaic solar cells. Solar Energy Materials and Solar Cells, 93(9), 1461-1470.

[62] Jin, S., Jun, G. H., Jeon, S., & Hong, S. H. (2016). Design and application of carbon nanomaterials for photoactive and charge transport layers in organic solar cells. Nano Convergence, 3(1), 1-10.

[63] Manzano-Ramírez, A., López-Naranjo, E. J., Soboyejo, W., Meas-Vong, Y., & Vilquin, B. (2015). A review on the efficiency of graphene-based BHJ organic solar cells. Journal of Nanomaterials, 2015.

[64] Das, S., Sudhagar, P., Kang, Y. S., & Choi, W. (2014). Graphene synthesis and application for solar cells. Journal of Materials Research, 29(3), 299-319.

[65] Alturaif, H. A., ALOthman, Z. A., Shapter, J. G., & Wabaidur, S. M. (2014). Use of carbon nanotubes (CNTs) with polymers in solar cells. Molecules, 19(11), 17329-17344.

[66] Wang, F., & Matsuda, K. (2019). Applications of carbon nanotubes in solar cells. In Nanocarbons for Energy Conversion: Supramolecular Approaches (pp. 497-536). Springer, Cham.

[67] Diener, M. D. (2018). Fullerenes for Bioscience & Photovoltaic Applications. URL: https://www.sigmaaldrich.com/technical-documents/articles/materials-science/nanomaterials/fullerenes. html (дата обращения: 20.10. 2018).

[68] Sood, A. K., Lund, I., Puri, Y. R., Efstathiadis, H., Haldar, P., Dhar, N. K.,... & Fritze, M. (2015). Review of graphene technology and its applications for electronic devices. InTech.

[69] Gupta, G., Zeng, M., Nurbawono, A., Huang, W., & Liang, G. (2016). 19 Applications of Graphene-Based Materials in Electronic Devices.

[70] Bharech, S., & Kumar, R. (2015). A review on the properties and

applications of graphene. J Mater Sci Mech Eng, 2(10), 70.

[71] Venkataraman, A., Amadi, E. V., Chen, Y., & Papadopoulos, C. (2019). Carbon nanotube assembly and integration for applications. Nanoscale research letters, 14(1), 1-47.

[72] Peng, L. M., Zhang, Z., & Wang, S. (2014). Carbon nanotube electronics: recent advances. Materials today, 17(9), 433-442.

[73] Mitra, M. (2018). Introduction on Carbon Nanotubes (CNT) and Its Applications in Electronic Circuits. Journal of Electronic Research and Application, 2(2).

[74] Paradise, M., & Goswami, T. (2007). Carbon nanotubes–production and industrial applications. Materials & design, 28(5), 1477-1489.

[75] Al-Mohamad, A., & Allaf, A. W. (1999). Fullerene-60 thin films for electronic applications. Synthetic metals, 104(1), 39-44.

[76] Papageorgiou, D. G., Kinloch, I. A., & Young, R. J. (2017). Mechanical properties of graphene and graphene-based nanocomposites. Progress in Materials Science, 90, 75-127.

[77] Aftab, S. M. A., Shaikh, R. B., Saifullah, B., Hussein, M. Z., & Ahmed, K. A. (2019, March). Aerospace applications of graphene nanomaterials. In AIP Conference Proceedings (Vol. 2083, No. 1, p. 030002). AIP Publishing LLC.

[78] Obodo, R. M., Ahmad, I., & Ezema, F. I. (2019). Introductory chapter: graphene and its applications. In Graphene and Its Derivatives-Synthesis and Applications. Intechopen.

[79] Wu, S., Qureshi, T., & Wang, G. (2021). Application of Graphene in Fiber-Reinforced Cementitious Composites: A Review. Energies, 14(15), 4614.

[80] An, L. B., Feng, L. J., & Lu, C. G. (2011). Mechanical properties and applications of carbon nanotubes. In Advanced Materials Research (Vol. 295, pp. 1516-1521). Trans Tech Publications Ltd.

Section 3

Electrochemistry of Advanced 2D Carbon Materials

Chapter 5

Electrochemical Exfoliation of 2D Advanced Carbon Derivatives

Muhammad Ikram, Ali Raza, Sarfraz Ali and Salamat Ali

Abstract

Advanced 2D carbon materials such as graphene and derivatives are basic building blocks for future nanostructured generation in electronics and energy horizons owing to their remarkable physical and chemical properties. In this context, production scalability of 2D materials having high purity with distinctive and multi-functionalities, that facilitate in fundamental research and advanced studies as well as in industrial applications. A variety of techniques have been employed to develop 2D advanced carbon materials, amongst state-of-the-art synthetic protocols, electrochemical is deliberated as a promising approach that provides high yield, great performance, low cost, and excellent up-scalability. Notably, playing with electrochemical parameters not only allows tunable properties but also enhances the content variety from graphene to a wide spectrum of 2D semiconductors. In this chapter, a succinct and comprehensive survey of recent progress in electrochemical exfoliation routes and presents the processing techniques, strategic design for exfoliations, mechanisms, and electrochemistry of graphene.

Keywords: 2D materials, electrochemistry, exfoliation, anodic exfoliation, cathodic exfoliation

1. Introduction

Two-dimensional (2D) materials motivated scientific society owing to inspired decisive passion in electrical, mechanical, and optical disciplines, showing extraordinary properties comparatively layered bulky counterpart. 2D pioneer carbon material, graphene, previously presented advanced studies in the fields, particularly, [1] membranes, [2] bio-sensors, [3] energy storage technologies, [4, 5] and topographic spintronics devices, [6] despite last decade advancement in graphene literature approach, still alarming goal from its targets, as is the condensed matter physics, [7, 8] towards the aforesaid trend, a series of ultrathin materials were isolated via exfoliation process, as synthesized incorporating metal chalcogenides, [9, 10] double-layered-hydroxide, [11] boron nitride, [12] preliminary investigation regarding 2D nano-materials was attractively oriented by fundamental research approaches inheriting novelty properties, new channels have certainly opened and encouraged recently towards high application inspired studies [13, 14]. Evidently, 2D materials frequently contributed active counterpart as a promising one in functional devices and versatile electronics. Eventually, they prove themselves as attracting candidates, revolutionizing the current technologies, further as, seawater desalination, quantum computing, and renewable energy resources [15–17].

Harvesting applications regarding 2D materials are expected to realize high efficiency with low-cost industrial-scale technologies should be appreciated in the development of high-quality 2D materials. Updates now reveal various top-down adopted methods, likewise, scotch-tape. Chemical and liquid-based exfoliation was followed, fabricating layered 2D materials successfully [18]. Recent investigations have shown remarkable information about top-down approach, regarding time-consuming, hazardous chemical nature, and more defects generation. Comparatively, epitaxial growth, and chemical vapor deposition (CVD), bottom-up approaches have considerable capability of fabricating ultrathin 2D materials containing large surface- area [19]. Nevertheless, aforesaid bottom-up methods are so complex that they show costly high temperature and pressure, rather, more need to transfer the 2D materials fabricated products from metal surface to targeted substrate, making difficult for controlling the synthesis process, and may incorporation of defects and impurities into the products. Electrochemical technologies are usually carried out under mild conditions, in comparison with, other synthesis technologies, as they proven convenient and controllable conditions [20, 21]. Electrochemical exfoliation, for the layered bulk-material, likewise, anodic-oxidation cationic-intercalation and cathodic-exfoliation, using liquid-electrolyte, applying potential driven structural expansion, is a potential method, exfoliating 2D materials in a remarkable novelty fashion [22, 23]. The electrochemical technique is also employed as a quick and controllable tool for lithium/non-lithium intercalations [15, 16, 24–31] and considered as an effective technique for exfoliating and/or intercalates layered carbon materials to single or multi-layered 2D nanosheets [32–35].

Electrochemical reactions occur on electrode with layered structure will yield as intercalation and/or exfoliation of electrode [36–38]. There are some desirable features for electrochemical exfoliation such as simplicity, fast cycle time, ease of activity, control, and potential for scaling up. The applied potential and electrolyte quality highly influenced on consistency of exfoliated nanosheets [39]. For this purpose, a set-up similar to the battery test system in a galvanostatic discharge mode with a constant current is used. In this context, a metallic lithium foil is used as anode and bulk Graphite powder is serves as cathode with $LiPF_6$ in a combination of ethylene carbonate and diethyl carbonate acting as electrolyte [16, 40]. Li^+ ions are introduced into graphene interlayer van der Waals gap during intercalation cycle and reduced by incoming electrons from the external circuit to Li atoms during insertion (**Figure 1**) [42]. Strongly in-plane covalently bonded bulk materials with weakly out-of-plane bonds, coupled by weak intermolecular forces, may easily be exfoliated in the form of thin-atomic layered structure of the 2D materials, by breaking weak van der Waals interactions under ultra-high cationic or anionic media [41, 43].

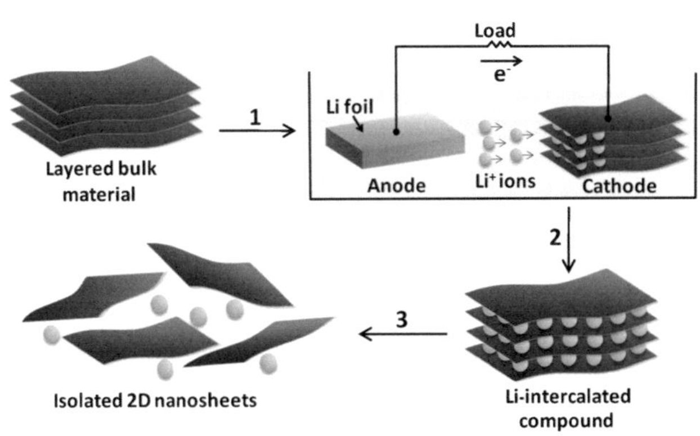

Figure 1.
Schematic illustration of electrochemical exfoliation [41].

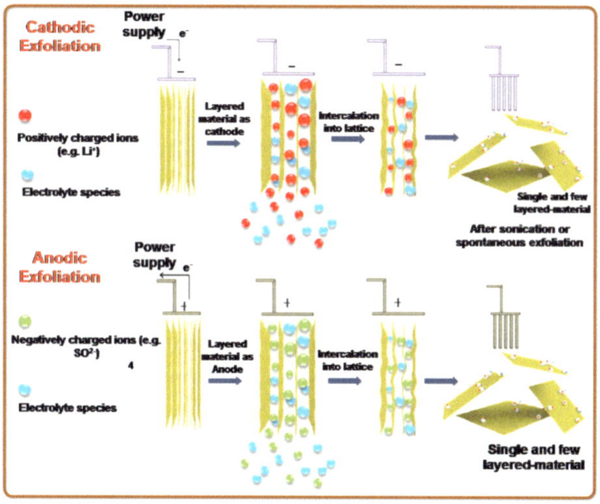

Figure 2.
Illustration of cathodic and anodic exfoliations.

The desired oxygen content, defect density, electrical conductivity, and thickness associated with exfoliated 2D materials, to be tuned, may be adjusted through voltage/current electrochemical parameters. Both cationic and anionic exfoliation, also intercalations, have been applied schematically in the exfoliation process of the graphite itself [44, 45], phosphorous black [46, 47] iv A and vA group metals [48, 49], transition-metal-dichalcogenides [32, 50, 51], graphitic-carbon-nitride, transition-metal-oxide [52], metal–organic-framework sheets [53] and MXene [54]. Based upon the type of potential used; electrochemical processes are mainly divided into two forms one is **(i) cathodic exfoliation**, performed in organic solvents such as Dimethyl sulfoxide (DMSO) and propylene carbonate comprising alkylammonium/lithium salts as electrolyte [16, 44, 55–59]. Other is **(ii) anodic exfoliation**, processed in ionic liquid or water mixtures or acids aqueous solutions such as H_2SO_4, $HClO_4$, H_3PO_4, and $H_2C_2O_4$; both exfoliations are described in **Figure 2** [22, 60–62].

2. History prospective of graphite intercalation chemistry

Graphite intercalation chemistry [63, 64] paves historical background path for the graphene, produced by electrochemically roots, the first step involves, typically, intercalation of ions in this respect. Scientists and engineers studied graphite intercalation compounds (GICs) over many decades, but exfoliation study of GIC was intensively increased to produce graphene/graphene-derivatives via characterization of graphene, employed by Geim and Novoselov [65]. A briefly reviewed of pre-graphene era work has been described here, included with the latest electrochemically produced graphene. GICs are identified, as numerous graphitic molecules resided between basic graphene sheets.. The intercalating molecules may play donor role in the graphitic network, otherwise, accept electrons (acting as accepters) to form chemically ionic-bond with graphite. Contrarily, a ternary GIC possibility prevails in the form of co-intercalated, acceptors and donors as well. GICs have interestingly presented considerable research study, owing to improved charming (electrical and electronic) properties relative to pure graphite. The very first reported literature on GIC was presented by Schafhäutl, in 1841 [66]. While, Various GICs methods have been promoted, producing the material under study, likewise, chemical photochemical and electrochemical synthetic approaches.

In addition, a homogeneous series of intercalating molecules were involved in various graphitic nature host materials [67], fabricating various GICs. GICs (amongst many species), including halogens, metal halides, alkali metals, and various acidic nature compounds are successfully incorporated into graphite. Electrochemical-intercalation-approaches have been studied since 1938, as Rüdorff and Hoffman employed electro-intercalation, to prepare acidic nature GICs [68]. However, until 1970s and 1980s, no interest has been taken in intensified electrochemically produced GICs. Moreover, in 1974, the Lithium/(CF) primary battery has been introduced by Fukuda while the 1970s presented the first lithium/graphite/fluoride battery-system on commercial basis, successfully [69, 70]. While electrochemical-intercalation approach was employed, here, a voltage is applied to graphitic working-electrode. In case the potential becomes positive, the graphite is identified as positively-charged anode, attracting anionic intercalating-species. In contrast, if the potential is opposite, then graphite acts as a negatively-charged cathode, which attracts cationic nature species. As a result, accordingly, both anionic-cationic intercalating-agents may be involved in the desired GICs. An anionic intercalating-species, which have been successfully incorporated, contained obviously sulfate- anions, fluoride-anions [71–73], and metal-halides respectively [74].

Cationic intercalating-species, including metals such as magnesium [34] and lithium have been reported [75, 76]. Lithium-ion GICs successfully exemplify the application of GICs towards the production of batteries, an area, where maximum research has been reproduced. GICs proved to be a successful battery cathode, or anode, or both alternatively. In the 1980s, lithium-ion GICs were progressed as anode-materials in secondary-batteries, associated with metal-oxide cathodes. Research into lithium-ion batteries progressively continues, currently, with due widespread commercial use this economical system. Furthermore, alternative GICs battery systems, such as metal-hydroxide-based systems [77], have also been adopted advanced steps and exhibited commercial based success. Various early electrochemically synthesized GICs products, based on the contemporary electrochemical-products of exfoliated-graphene and functionalized-graphene, i.e. early work on lithium/GICs advanced materials, which would be exfoliated to graphene, later on, were also appreciated [78]. Stage-I, earlier GIC literature on GICs, is considered the most relevant current-work on graphene exfoliation approach. As far as Stage I is concerned, compound is formed during the process of one layer of graphene resided between every layer of intercalating-molecules, whereas Stage-II GIC shows two-layers of graphene intercalated between each layer of guest-molecules. Stage-III GIC contains three-layered groups of graphene residing guest molecules, and continue simultaneously. Since Stage-I GICs, the guest species, enlarge the inter-layer spacing between graphene layers, following basic principle, each layer may easily be separated from its neighbor one, so becoming able to be exfoliated into single-layered graphitic nature. Much electrochemical-graphene work, decisively first creates Stage-I GICs, which are, later on, exfoliated in the form of monolayers. Earlier study reveals that electrochemically produced Stage-I GICs have been announced more informative in many studies, clearly described in the forthcoming sections. It is very likely, and innovatively, that this literature study will continue to be made a foundation for future work, successfully [79].

3. Electrochemical setup and exfoliation mechanisms

3.1 Experimental setup

The electrochemical setup, used for graphene exfoliation, usually incorporates the elements such as graphite working-electrode, counter-electrode, reference-electrode, electrolyte, and voltage-supply. Systematically, highly-orientated pyrolytic-graphite (HOPG), graphite-powders, graphite-rods, graphite-foil, or graphite-flakes has been used as the working-electrode [22, 44, 80]. To provide the conducting surface, graphite flakes were choosed as the best, amongst available electrodes, that may be adhered to conductive carbon-tapes, forming the working-electrode [22, 45], and they may also adhere to tungsten-wire via silver-pad [81] or to be formed into graphite-plates through compression directly [82]. Being counter electrodes mesh, platinum-wire, plates or rods, and graphite were more frequently used. The arranged experimental setup is often illustrated as depicted in **Figure 3a**. Keeping a certain distance between working and counter electrodes respectively, they are simultaneously immersed into electrolyte. A voltage (positive or negative) is applied to the graphite (a working electrode), depending upon adopted desired exfoliation mechanism.

In addition to the aforesaid common setup, Liu et al. employed two pencil cores, as graphitic anode and cathode sources alternatively [80]. An alternating bias-voltage (between +7 V and − 7 V) was applied across the ends of pencil-electrodes, exfoliating them properly. Though the setup was highly efficient with higher exfoliation rate than graphite electrode, yet the product so obtained may be expected more inhomogeneous, with wide thickness and suitable size distribution. Abdelkader et al. reported, recently, a versatile setup in **Figure 3b**, showing continuous electrochemical-exfoliation-process, producing 0.5–2 g (few-layer graphene) per hour [83]. Moreover, in the setup, the graphitic electrode was injected steadily from the bottom of the electrolytic cell with graphitic contact with the electrolyte, being so exfoliated. Well- immersed-exfoliated (few-layer graphene sheets) was located on upper surface of the electrolyte, thereby, flowed out of the cell, while the partially-exfoliated-graphite retained at the bottom, so that further exfoliation may be carried out [83]. In another study, Motta and coworkers have presented ultra-sonication, assisting the electrochemical-exfoliation process, and placing the graphite electrode in a sonicated-exfoliated process [84].

Sorokina et al. introduced a patent experimental setup, comparatively, producing GICs in the past of the graphene era indicating a load (20 kPa) was applied across graphite-flakes over a platinum-disk (electrode), so to achieve fine electrical-contacts between the graphite-flakes as well [85]. Recently, the main challenging issue lies between (the effective and uninterrupted) electrical-delivery, to each graphene layer, in the graphite, presenting the immense need for the development of commercially scalable, and further controllable-setup.

3.2 Electrode preparation

Various bulk-layered materials exhibit strong in-plane bonds while electrostatic interactions with weak interlayer bonding i.e., interlayer-cohesive-energies (less than 200 meV/atom) [18]. So, exfoliation or delamination occurred in the form of atomically thin-layered nanosheets, thereby, van der Waals forces amongst 2D binding layers reduce to a minimum level. Mechanical exfoliation/chemical exfoliation as compared with ultrasonic treated exfoliation was extensively carried out fallowed by two-electrode or three-electrode electrolysis of electrochemical exfoliation (using bulk-material as working-electrode). Plasma state as well as cations or anions accumulated between layers owing to a strong electric field, resulting in layered-structure electrodes expansion with the interlayer-bonding cleavage simultaneously. Hence, bulk-layered-structured material may prove to be a good conductor of electricity, thereby, could be made electrode. It has been reported that bulk layered materials are semiconductive as well as non-conductive in nature [86] caused by difficult to be electrochemically exfoliated, as in this case, the most applied potential causes overwhelming large resistance. To overcome issue, a

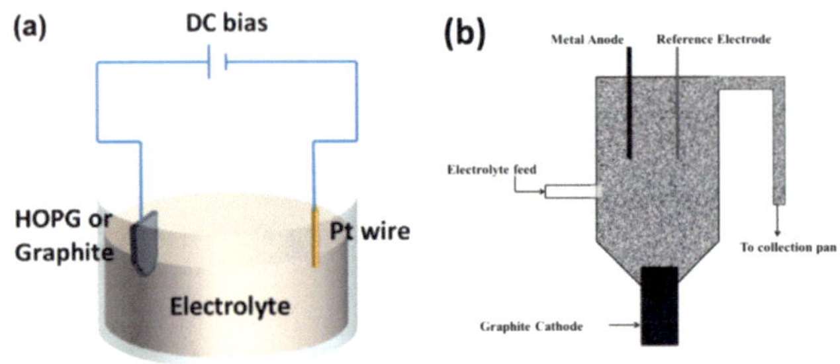

Figure 3.
(a) Schematic illustration of a typical setup for electrochemical exfoliation of graphite [81], (b) schematic of the electrochemical cell for continuous process [83].

conductive additive is suggested to be more appropriate strategy [61], resulting in exfoliation of 2D layered materials in an extensive range of possibility while ignoring conductivity of the bulked layered materials.

During the exfoliation mechanism, expansion of bulk material electrode occurs under the intercalation of ions, leading to disintegration of bulk material electrodes. Resultantly, some disintegrated sheets were still not exfoliated, reducing the yield strength and preventing electrochemical exfoliation process from the possibility of feasible production route. During the intercalation process, chances of breaking of bulk material electrodes, they are wrapped up in confined space with plastic tube and platinum gauze or carbon cloth, suggesting reasonable method for laboratory preparation method [87, 88]. Currently, Achee et al. framed a new route, yielding highly scalable 2D graphene by employing graphite flakes, without binder as the working electrode [89]. Graphite flakes remained in electrical contact under the compressed expandable electrode system, expanded by gas evolution. Therefore, graphene powders accumulated continuously expanded largely, and exfoliated extensively to produce carbon materials (graphene), 2D in nature.

3.3 Electrochemistry of exfoliated graphene and mechanism

The electrochemical exfoliation Mechanism depends on the type of applied potentials (anodic or Cathodic, **Figure 4**). Amongst the going mechanisms, anodic-exfoliation contains an anionic-intercalation with any co-intercalating-species (in the reaction mixture) into graphitic nature material. A positive current extracts electrons from the graphite (a working anode), thereby producing a positive charge. The charge, so produced, proceeds of bulky negative ion's intercalation like sulfate anions, that have increased the interlayer-spacing between graphene-sheets, and further supported during the exfoliation of the sheets, subsequently. A negative biased graphitic working-electrode in cathodic exfoliation attracts positively-charged-ions (e.g. Li^+) in the electrolytic solution, involving any co-intercalating molecules. Furthermore, the intercalating species create a location where they open the graphene sheets, depending upon expansion and exfoliation processes [16, 90, 91].

After completion of electrochemical intercalation along with expansion of graphite, further need is required to some form of exfoliation. In some cases, where exfoliation-process may occur during which intercalates (more typically), or the co-intercalating species, such as water, that was rapidly transformed to expanded-species (e.g. oxygen gas) [81]. On the other hand, electrochemically expanded graphitic sheets requires, to be mechanically-exfoliated likewise sonication process [78]. The exact mechanism related to electrochemical-graphene-exfoliation

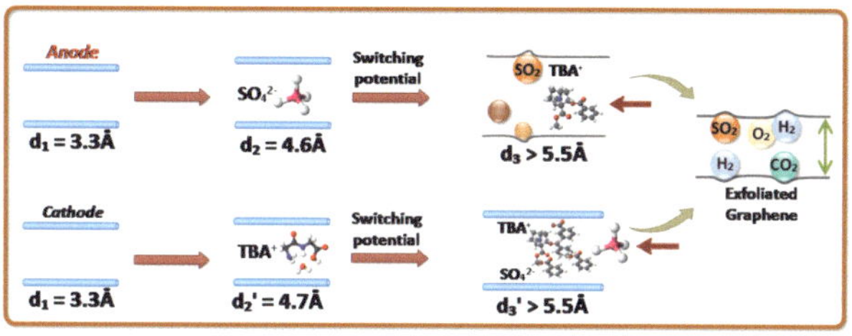

Figure 4.
Proposed mechanism for exfoliation process at both anode and cathode.

depends upon the potential polarity, along with other experimental conditions, caused by the electrolyte as well as co-intercalating agents already incorporated in the mechanism, to be further discussable (vide infra) [15].

An anodic exfoliation mechanism in ammonium sulfate ($(NH_4)_2SO_4$) aqueous solution, outlined by Parvez et al. [22]. In (**Figure 5a**), Hydroxyl ions (OH^-), firstly produced from the water electrolysis, and this strong nucleophile may interact the sp^2 carbons graphitic- edges with grain boundaries, thus producing two vicinal hydroxyls (OH) groups. Subsequently, they interact with each other, exploring epoxide group rings. Alternatively, dissociating them forming of two carbonyl-groups via further additional oxidation, as illustrated in **Figure 5b**, reaction (3). Resultantly, this leads to depolarization with an expansion of graphitic-layers at the corners, which in turn opened up the lattice, for intercalation, by sulfate ions SO_4^{2-}, providing opportunity towards possibly more water molecules. In addition, along with the oxidation of graphite, further reactions are certainly expected to occur, such as involvement of evolution of (CO_2 and O_2 gases respectively) by performing reactions 4 and 5 in **Figure 5**. CO_2 and O_2 gases also assisted reasonably during the exfoliation of the graphitic layers [22].

Similarly, anodic process was also described by Rao et al. [92]. Hydroxyl ions (i.e. OH^- ions) from aqueous NaOH electrolytic solution reacted with more added H_2O_2 to form O_2^{2-} ions that have proved to be more nucleophile than OH^- ions. That is why, they may be easily intercalated into graphene-sheets, with the aid of (a positive) electrochemical-potential. As an example of a cathodic exfoliation mechanism, Li^+ (positive ions) in organic solvent PC (propylene carbonate) may be systematically used as intercalating-agents [44, 78]. Electrochemical process was achieved by the co-intercalation of PC and Li^+ ions in the form of negatively charged graphitic layers, as illustrated in **Figure 6**.

By supplying sufficiently high voltage, the organic solvent will be decomposed, producing propylene gas which added the graphitic expansion [44].

Alkaline situations along with 1 M of sodium hydroxide (NaOH) and father explore the impact of adding hydrogen peroxide (H_2O_2) on exfoliation efficacy, experimental setup with mechanism as shown in **Figure 7a, bi-ii**. The existence of H_2O_2 considerably improves the exfoliation due to formation of extremely nucleophilic ions (O_2^{2-}) that causes to intercalate and magnify graphene layers. This corresponds to the extremely reactive radicals (i.e. O and OH) produced by firstly, anodic oxidation of water and secondly, opened and oxidized the edge sheets assisting intercalations of the peroxide ions **(Figure 7f-g)**. The exfoliation route happens tremendously fast and obtained graphene sheets attaining a low density of defects and low oxygen group content **(Figure 7c-e)**. Further, exfoliation approaches for graphite using anodic mechanism were projected using phosphate, nitrate, carboxylate, and perchloride [16, 93]. Likewise, Abdelkader et al. used Li^+ and alkylammonium ions (Et_3NH^+), in dimethyl sulfoxide (DMSO), intercalating into graphitic-layers, while weakening the van der Waals interactions between the

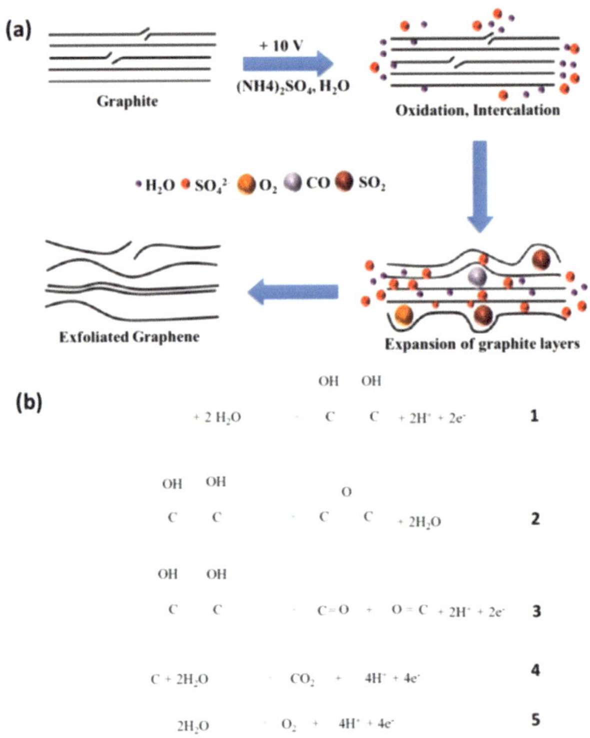

Figure 5.
(a) Schematic illustration of mechanism of electrochemical exfoliation in $(NH_4)_2SO_4$ aqueous solution [22], (b) electrochemical oxidation reactions occurs at anode for graphite exfoliation [56].

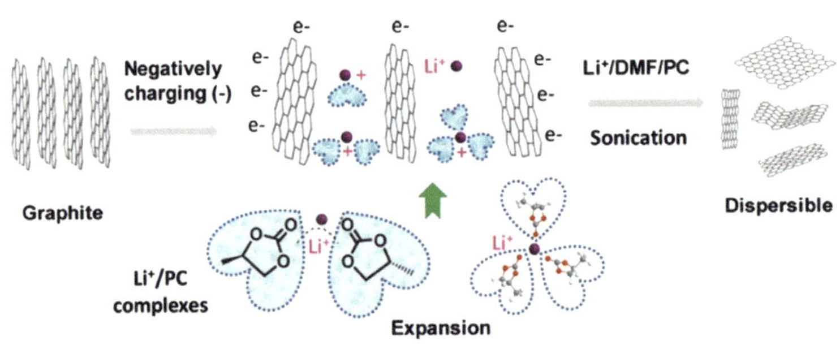

Figure 6.
Exfoliation of graphite into few-layer graphene flakes via intercalation of Li^+ complexes [44].

layers [83]. Simultaneously, Et_3NH^+ was likely reduced electrochemically to Et_3N gas, supported by graphitic exfoliation successfully.

3.4 Anodic exfoliation

Amongst many electrochemical exfoliation methods, anodic graphite exfoliation is that one, showing high exfoliation efficiency. Various diversified graphene production approaches were adopted, based on anodic exfoliation, which has already been reported [22, 60, 94–96]. Su et al. presented the best one approach (as the first reported) of anodic exfoliation, via adopting the most simple and fast method, while preparing electrolyte solution containing H_2SO_4 + KOH [81]. An optimized procedure that was followed here, for the exfoliated graphene production was the setup, similar to what is shown in **Figure 3a**, using the electrolyte with value (pH = 1.2). A low-biased +2.5 V has been first applied for 1 min, yet with

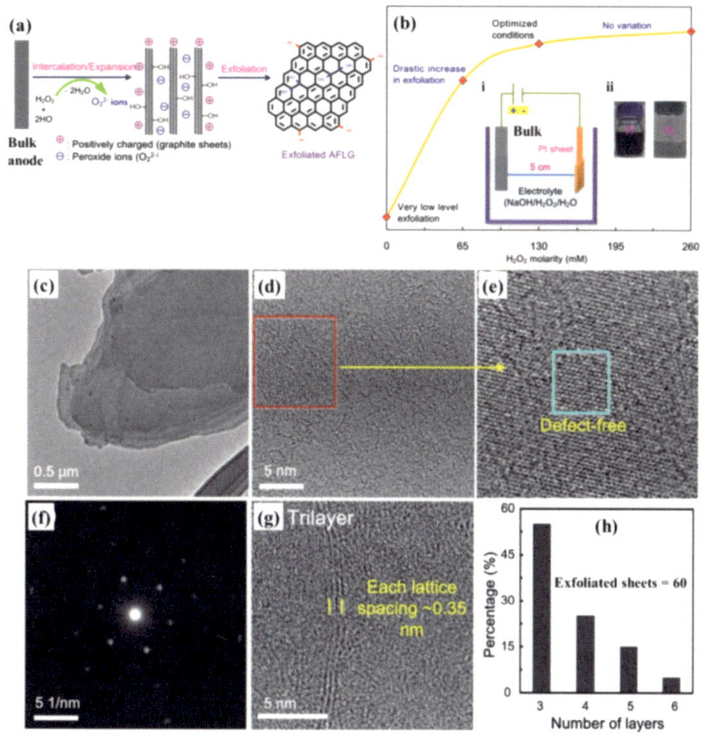

Figure 7.
(a) Schematics of proposed mechanism of anodic exfoliation (bi, ii) experimental setup and exfoliation efficiency against H_2O_2 molarity with photograph of dispersed nanosheets in C_3H_7NO. (c, and d) low magnification (0.5 μm) and HR-TEM images of exfoliated nanosheets, respectively, (e) image reveals some defects in nanosheets (f) SAED image (g) HR-TEM image, exposing tri-layer formation, (h) distribution of exfoliated nanosheets before centrifugation [92].

subsequent alternating-voltage between +10 V and – 10 V. In first step, low-voltage, aided for forming the wetting electrode surface, helping intercalation of anions into the graphite. Subsequently, the +10 V potential was used, for activating and oxidizing the graphitic sheets, which caused the graphite to become quickly in the form of dissociated small pieces. The ensuing (−10 V) potential was used as reductants towards functional groups. Very impressively, the so produced graphene sheets show a lateral size of several to 30 μm. Above 60% of the sheets were observed as bilayer-graphene with A–B stacking as illustrated in (**Figure 8**). Oxygen functional groups along with some decisive defects have been detected in the graphene sheets attributing to unavoidable oxidation. Moreover, the concentration level of graphitic defects produced in graphene sheets was less than reduced graphene oxide, which was produced by traditional chemical methods.

A similar study was presented by Su and colleagues [81], showing optimized multiple parameters, involving pH as well as applied voltage. While at extremely low pH, with high oxidation levels including H_2SO_4, produced a maximum level of defects on the graphene sheets. Consequently, KOH was added along with H_2SO_4, to increasing the pH value of the electrolyte, exhibiting the exfoliation at lower rate. Resultantly, it was observed that higher concentrated pH showed large percentage of bilayer-sheets, but the non-uniform defect level was still maintained between the graphene sheets. Subsequently, at less than 10 V potential (in terms of the working biased potential), the exfoliation process was slowed down and more inefficient, whereas voltages (greater than 10 V) accelerated the exfoliation rate very fast so that density of graphitic-particles, as well as, thickest graphene sheets were clearly observed and largely produced. Obviously, the effects of various electrolytic

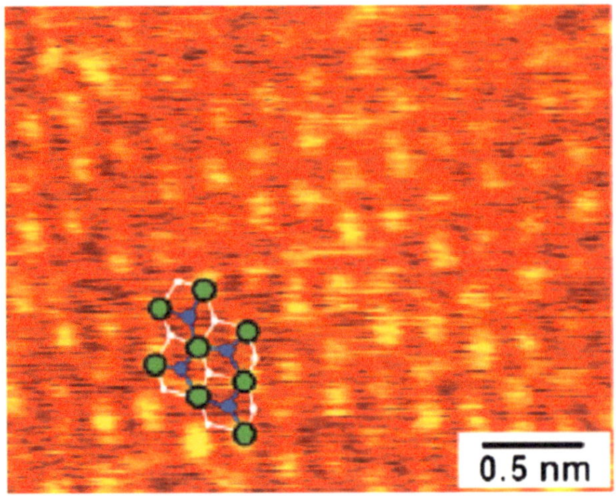

Figure 8.
STM image of bilayer graphene produced by Su et al. hexagons represent atom configuration of two layers [81].

solutions were greatly explored, involving some acids, such as HBr, HNO_3, HCl, and H_2SO_4, however, amongst the aforesaid solutions, H_2SO_4 was found only to be more effective in the performed experiments.

In 2013, Parvez et al. contributed and demonstrated their work in the form of exfoliation process of graphite in H_2SO_4 aqueous solution, further proceeding and elucidating, the exfoliation mechanism as well [45]. In this respected end, they have been explored the influence of H_2SO_4 concentration more clearly on exfoliation performance, by using (+10 V voltage), for 2 minutes subsequently. It was, more certainly, found that 1 M and 5 M H_2SO_4 explored slow exfoliation efficiency and yielded 0.1 M H_2SO_4, presumably, because of (more concentrated H_2SO_4 solutions), generated larger fragments of graphitic-particles. Likewise, in case of sulfuric acid, was too low, the exfoliation efficiency was more frequently reduced, caused by a reduced number of anions. The worthy authors have deeply studied while examining pure H_2SO_4 with 1:1 H_2SO_4/CH_3COOH reaction mixture, however, in these cases, slight expansion with almost no exfoliation was prominently observed so far. This scheme has suggested the durability of water in the electrochemical process, as it clearly may produce (oxygen and hydroxyl radicals), which arises as aiding agents in intercalation and exfoliation processes. High-quality graphene was exfoliated via 0.1 M sulfuric acid solution, with a large sheet, containing a size of ~10 μm, with low oxygen concentration 7.5 wt.% along with low sheet-resistance (of 4.8 kΩ/square), for a single sheet as in **Figure 9a-f**.

Liu et al. presented electrochemically exfoliation of two graphitic-electrodes, through applied alternating potentials (+7 V and − 7 V) in aqueous electrolytes, containing H_2SO_4 or H_3PO_4, thereby, resulting in anodic-exfoliation using both electrodes alternately [80]. Depending upon Characterization results, graphene flakes with thick multilayered structure (3–9 nm), lateral size (1–5 μm) with comparatively low oxidation level, were produced (see **Figure 10**).

Xia et al. keenly observed, the swallowed and expanded graphitic surface, caused by the intercalation along with gas formation at early stage level [74]. Apparently, opening of graphitic edges is caused by a key-step towards the subsequent exfoliation. Furthermore, the radical attack was observed as nonselective, in this case, occurring randomly at the exposed graphitic surfaces, necessarily leading to increased oxidation level of the graphene sheets. Partial removal of the radicals indicates a sound solution, preventing the side reaction, so occurred.

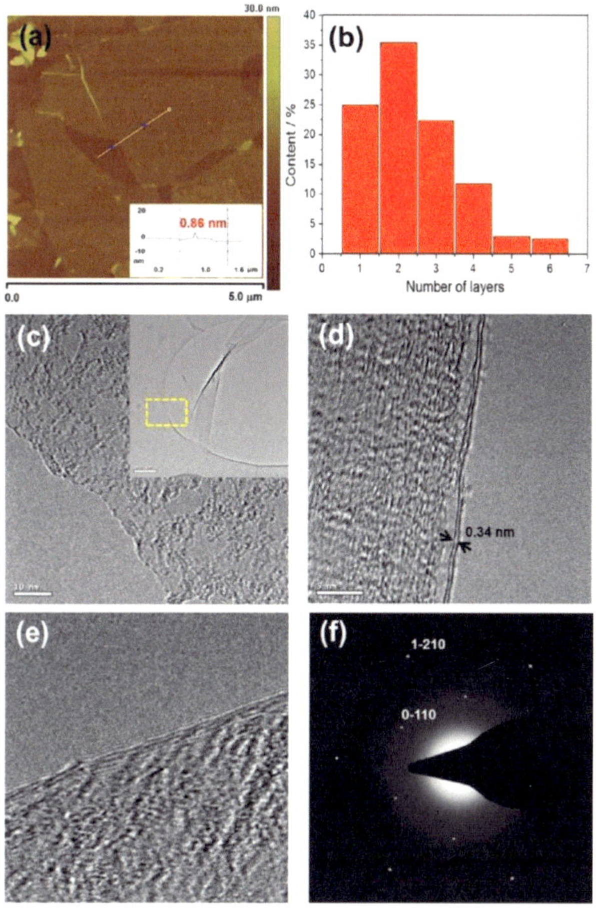

Figure 9.
(a) AFM image of electrochemically exfoliated graphene on substrate (SiO_2), (b) statistical thickness analysis of the graphene sheets by AFM, (c, d, and e) HR-TEM images of single-, bi-, and four-layer graphene; inset in (c) is the low magnification image of exfoliated graphene, and (f) SAED pattern of bilayer graphene [45].

Yang et al. [97] have examined an antioxidants group, based on a standard ammonium sulfate $(NH_4)_2SO_4$ electrolyte, and with radical scavengers containing sodium borohydrides, ascorbic acid, (2,2,6,6,tetramethyl-piperidinyl)oxyl (TEMPO) acting as additives candidates during the exfoliation process. Consequently, the more addition of TEMPO causes greatly suppressed oxidation state, yet not compromised the exfoliation efficiency, with production of 15 g $h-{}^{1}$ showing high quality graphene, exploring large dimensions (5–10 μm), but only few defects were observed in the form of C/O ratio equal to 25.3. **Figure 11** showed that TEMPO initially reacted with the ($HO^{\bullet}$ radicals) at anodic end, generating metastable TEMPOOH along with oxoammonium cations. At the Cathodic end, the aforesaid intermediates (compounds) were largely reduced to TEMPO radicals in again turn. In the system discussed here, single graphene sheets appeared to be an ultrahigh hole-mobility upto 405 $cm^2 V^{-1} s^{-1}$, owing to be still an excellent processibility in N,Ndimethylformamide (DMF) (6.0 mg mL^{-1}), preparing graphene ink as well (**Table 1**).

3.5 Cathodic exfoliation

For decades, a graphitic negative electrode has been extensively used in lithium-ion battery-technology, owing to its high electrical conductivity and ability,

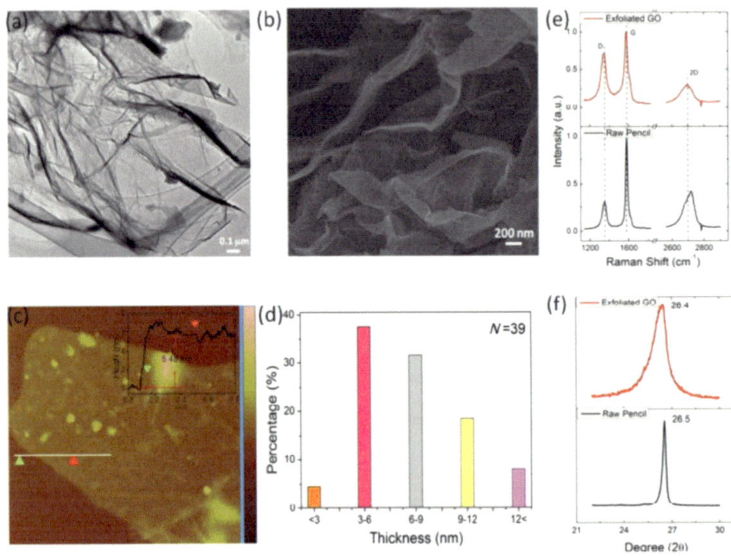

Figure 10.
(a) TEM image and (b) SEM image of exfoliated GO flakes, (c) AFM image of exfoliated GO flakes. The thickness is 5.45 nm with lateral size around 2 µm, (d) thickness distribution histograms for exfoliated GO sheets, as estimated from corresponding AFM analysis. The graphene flakes are mainly distributed in the range of 3–9 nm thickness (69%) with lateral size about 1 to few µm, (e) Raman spectra, and (f) XRD patterns for both pencil core and exfoliated GO flakes, respectively [80].

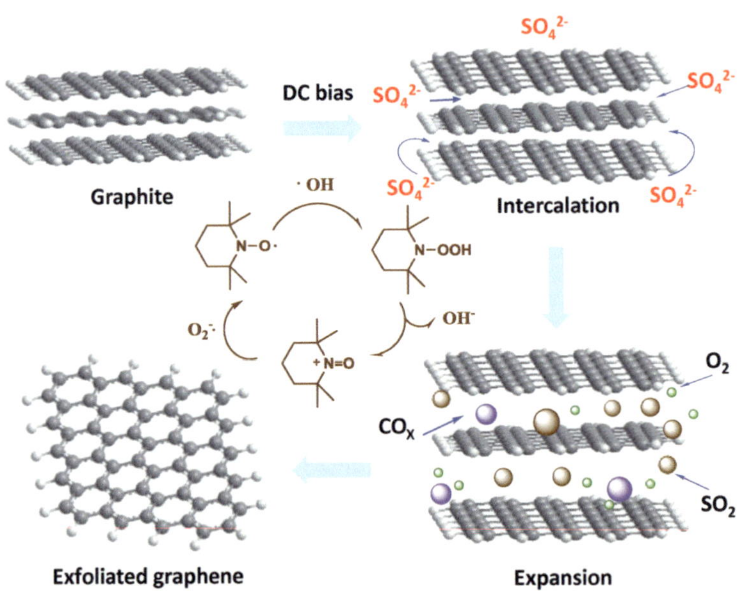

Figure 11.
Anodic exfoliation of graphite in an aqueous electrolyte with sulfate anions and TEMPO. TEMPO is a radical scavenger that partially eliminates the hydroxyl radicals from water oxidation [97].

for hosting lithium between the graphitic layers (**Figure 12**). In this way, the lithium-graphitic intercalation-compounds decomposed into water at a very fast rate, giving rise to lithium hydroxide along with free-standing graphene sheets. The aforesaid principle has been recently introduced, as a durable route towards scalable production of graphene [107]. However, depending on slow kinetics of the intercalation-process, the lithium was bounded to those areas closed to the edges. Upon exfoliation into water, graphitic expanded edges were clearly produced

Bulk materials	Electrolytes	Working potentials	Yield	Thickness	I_D/I_G	Ref.
HOPG/natural graphite	0.5 M H_2SO_4 + KOH (pH ≈ 1.2)	1) +2.5 V, 1 min; 2) Switching +10 V, 2 s; −10 V, 5 s	5–8 wt%	≤2 nm	0.5–1.0	[81]
Graphite foil	0.1 M H_2SO_4	+10 V, 10 min	60 wt%; 4.2 g h^{-1}	1–3 layers	0.4	[45]
Graphite foil	0.1 M $(NH_4)_2SO_4$	+10 V, 10 min	75 wt%; 16.3 g h^{-1}	1–3 layers	0.25	[22]
Graphite foil	0.05 M NaCl	+10 V, 60 min	—	2–3 nm	0.8	[98]
Expanded graphite foil	0.1 M NaOH + Na_2SO_4	1) +3 V, 3 min 2) +10 V, 30 min	—	2–3 nm	1.3	[99]
Expanded graphite foils	0.1 M $(NH_4)_2SO_4$ + 1 mg mL^{-1} TEMPO	+10 V, 10 min	75 wt%; 15.1 g h^{-1}	1–3 layers	0.1	[97]
Expanded graphite foil	0.5 M Na_2SO_4 + 0.05 M $CoSO_4$	+20 V, 120 min	—	Monolayer and few layers	0.05	[100]
Bulk graphite/ graphite powder	0.1 M H_2SO_4 + 1 mg mL^{-1} melamine	±20 V, 10 min	1.5 g h^{-1}	1–3 layers	<0.45	[101]
Graphite foil	0.2 M SNDS in water	+10 V, 60 min	—	2.5 nm	0.2	[36]
Graphite foil	30 mg mL^{-1} $LiClO_4$ in PC	−15 ± 5 V	>70 wt%; 0.12 g h^{-1}	<5 layers	<0.1	[44]
Graphite foils	0.1 M TBA HSO_4 + NaOH	±10 V, 0.1 Hz	75 wt%; 20 g h^{-1}	1–3 layers	0.15	[102]
Graphite foils	0.1 M $(NH_4)_2SO_4$ + 1% thiourea	±10 V, 0.1 Hz, 60 min	—	—	0.06–0.14	[103]
HOPG/graphite rod	0.1 M $(NH_4)_2SO_4$	switching +7 V or + 10 V, 1 s; −0.5 V, 3 s	77 wt%	<5 layers	0.29	[34]
Graphite foil	0.5 M $LiClO_4$ in water	1) +2.0 V, 2 min 2) +10 V	– (graphene oxide)	6–8 layers	1.0	[104]
Graphite flakes	1.0 M H_2SO_4 in saturated $(NH_4)_2SO_4$	anodic, 0.6 A, 24 h	40 wt% (graphene oxide)	1.5 nm	1.0	[55]
Graphite foil	1) 95% H_2SO_4 2) 0.1 m $(NH_4)_2SO_4$	1) +2.2 V, 10 min 2) +10 V	71 wt% (graphene oxide)	monolayers	1.48 ± 0.01	[105]
Graphite foil	1) 98% H_2SO_4 2) 50% H_2SO_4	1) +1.6 V, 20 min 2) +5 V	96 wt%; 12 g h^{-1}	1–3 layers	>1.0	[106]

Table 1.
A summary of electrochemical exfoliation and anodic oxidation of graphite.

and further intercalation also occurred positively, thereby, also water decomposition and sonication steps were necessarily taken, achieving complete exfoliation (**Figure 13**) [44]. Liu et al. and Huang et al. [108, 109] have presented much effort, for accelerating the intercalation kinetics, by using molten (LiOH or LiCl) at 600°C. However, the intercalation was considered there so insufficient to be achieved perfect graphitic exfoliation, but sonication steps thus fallowed were still required to achieve remarkable production of graphene-based materials.

Swager and Zhong [78] suggested a synergetic method to be intercalated the graphite primarily with Li^+, by following tetra-alkyl-ammonium cations into two steps separately. Moreover, due to expanded nature of the cathode, the distance between electrodes was kept initially very large, exploring the high potential difference required to apply, to dominate the high Ohmic-drop, created by the electrolytic cell configuration. Resultantly, the organic electrolytic solvent was dissociated in that state, occurring later on, at all the stages of the procedure by disappearing slowly during intercalation process. That is why, additional steps were rendered through sonication mechanism again, need to be sufficient for achieving reasonable exfoliation proceedings.

Dimethylsulfoxide (DMSO) has shown a wide electrochemical window that is highly efficient solvent during the graphene solvent dispersion, reflecting typical dispersive qualities, by comparing those of NMP [110]. As a result; DMSO forms various solvated ions, containing both lithium and alkylammonium ions reasonably. The observed solvated ions are expected to be able to intercalate with graphite, via decomposition between the graphene layers making SO_2 and/or along with amine-based apparent gases. The stress applied properly on the graphene sheets through the gaseous expulsion so occurred is evaluated enough to overcome the forces (van der Waals) that attracting the neighboring sheets, allowing separation of graphene sheets formed by the graphitic cathode, thereby, allowing dispersal occurring in the electrolytic solution. The authors of the literature [83] have applied the said principle to make many flakes, showing lateral dimension (upto 20 μm) of few-layer graphene towards DMSO-based electrolytic solution, containing triethylammonium and Lithium ions. Authors have adopted fashioned of electrochemical program, by applying a controlled Cathodic-potential towards the graphitic electrode, which presents complete intercalation prior to flakes formation spontaneously, so that exfoliation from the Cathodic end because of partial expansion occur consequently. It was greatly suggested that the triethylammonium ions, dissociated between the layers, give rise to triethylamine along with hydrogen gases, highly encouraging the exfoliation of flakes more prominently.

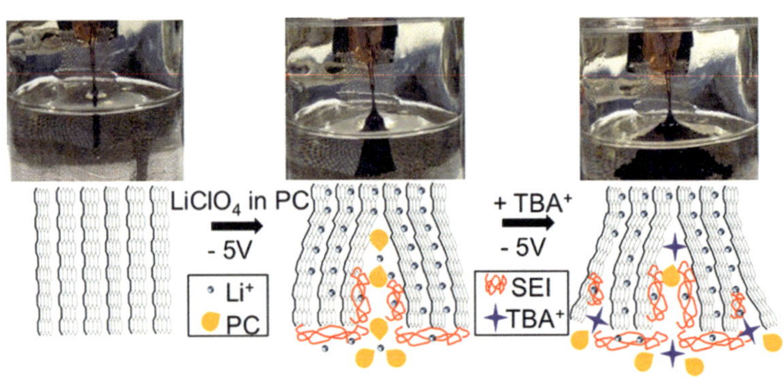

Figure 12.
Schematic and images of cathodic electrochemical expansion of graphite.

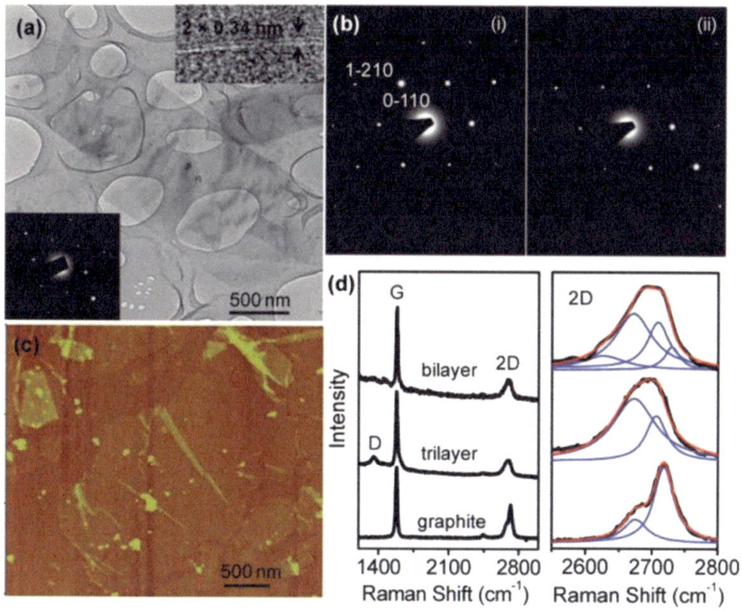

Figure 13.
(a) TEM images and electron diffraction pattern of cathodic exfoliated graphene, (b) electron diffraction patterns of (i) single and (ii) bilayer sheets, (c) AFM image of exfoliated graphene spin-coated onto a Si substrate. The thickness is ~1.5 nm, corresponding to a bilayer. (d) (left) Raman spectra (532 nm laser) on Si substrates compared with the spectrum of graphite; (right) Lorentzian peak fitting of the 2D bands of the bilayer and trilayer [44].

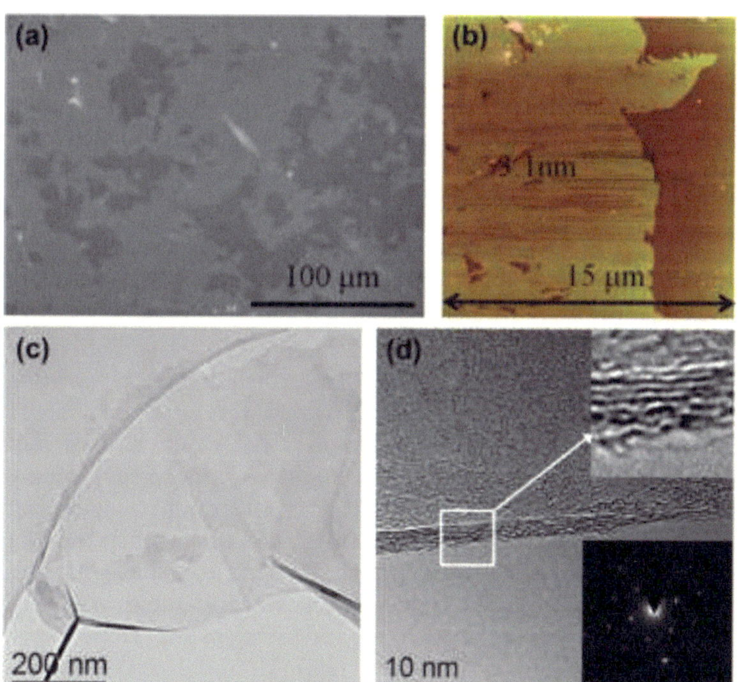

Figure 14.
(a) SEM image, (b) AFM image of graphene flakes deposited on Si substrate, (c) TEM image, and (d) HR-TEM image of a graphene flake. The inset is an electron diffraction pattern and magnified portion of the edge of the graphene flake [111].

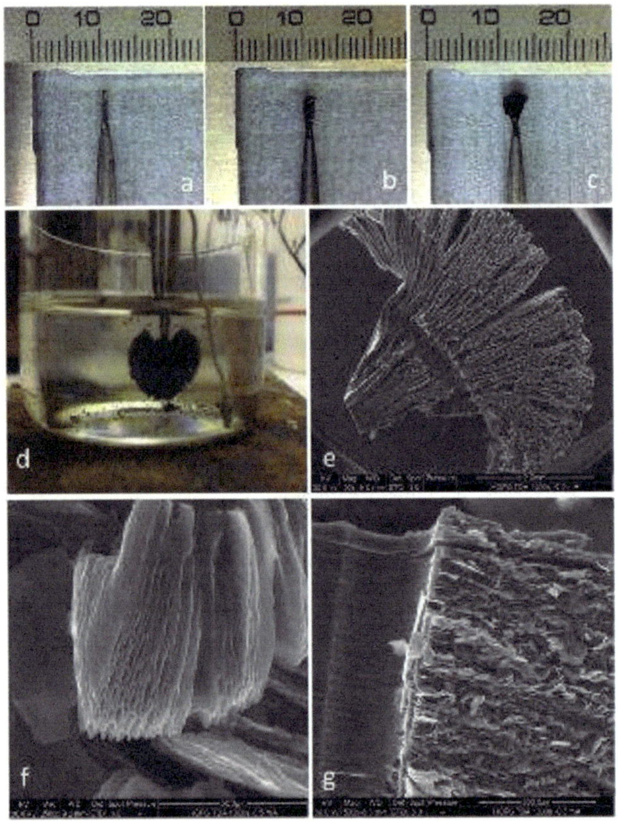

Figure 15.
(a) Photographs of as prepared HOPG, (b) HOPG expansion after 1000 s tetraethylammonium cation intercalation, (c) HOPG expansion after 1000 s tetrabutylammonium cation (TBA^+) intercalation, (a–c scale in mm) (d) HOPG expansion after 10,000 s TBA^+ intercalation, (e) SEM image of HOPG expansion after 6000 s TBA^+ intercalation, (f) SEM image showing micron-sized pores in HOPG after TBA^+ intercalation, (g) SEM image showing selective exfoliation of HOPG electrode: The point on a HOPG electrode that was held by tweezers (left-hand side) whilst the rest of the electrode (right-hand side) was submerged [57].

Zhou et al. [111] have efficiently presented, so far, the only familiar method followed to exfoliate graphitic cathodes into aqueous medium deliberately, using an electrolyte containing NaCl, DMSO, and thionin acetate salt. Sodium ions were chemically combined with (four or five) DMSO molecules, readily forming Na^+/DMSO complex-composite. Complexes so obtained were still intercalated in the form of graphene-galleries owing to graphite, clearly forming ternary graphitic-intercalation compounds ($Na^+(DMSO)_yCn^-$). Further, interlayer spacing was systematically reported to be 1.246 nm, accordingly. However, perfect exfoliation was rather not achieved through only electrochemical-treatment, therefore the sample was necessarily subjected to sonication process in order to achieve more stable graphene dispersions (**Figure 14**). In addition, however, samples were observed as heavily contaminated (with sulfur, oxygen, and nitrogen impurities).

Cooper et al. have deliberately shown tetraalkylammonium salts to be cathodic intercalation into HOPG by using relatively low potentials (ca. −2 V) [112] and maybe systematically employed to produce purely cathodic-exfoliated materials, consisting clearly (2 or 5 layers) of graphenes (see **Figure 15**) [57]. More significantly, the materials were certainly found containing (no functionality or oxidation), rather inclusion of slightly 1% in atomic form oxygen, probably induced from the atmospheric exposure of the so obtained isolated materials.

Further, Yang et al. [113] employed a pure ionic-liquid, N-butyl, methylpyrrolidinium bis (trifluoromethylsulfonyl)-imide (BMP TF_2N) towards cathodic-graphitic intercalation/exfoliation mechanism. In authors' view, $[BMP]^+$ cations chemically intercalated between the highly negatively charged (graphene layers), causing the expanded interlayer spacing. The aforesaid expansion facilitates the bigger molecules insertion, such as the $BMPTF_2N$ ion-pair, subsequently, caused by higher expansion in graphite as well. The authors have certainly claimed that formation of graphene sheets was consisted of between (two and five layers), with 2.5% atomic-oxygen yet free defected materials. However, the authors, not suggested a reasonable explanation for the gel-like-phase, probably formed from the ionic- liquid during which (the cations or anions) are expected to be consumed in all irreversible reactions [114, 115].

4. Conclusions

The process of electrochemical exfoliation has been confirmed to operate in a wide variety of layered materials; the majority of studies are conducted on large-sized bulk single-crystals, which are costly and inefficient for industrial applications. Small-sized powders or flakes are readily produced from natural materials or industrial synthesis should be considered as an alternative for efficient and successive exfoliation. Both aqueous and non-aqueous electrolytes are employed to exfoliate layered materials, but the procedure is more often used in aqueous solutions and under anodic conditions for the exfoliation of graphite owing to better performance relative to cathodic scheme, in this technique most reliable and effective way is Li-ion insertion. Around the same time, a deeper understanding of process/mechanism of intercalation and exfoliation of powered by application of current is desperately required, which may encourage the use of electrochemical means to exfoliate more effectively a large number of layered materials.

Conflict of interest

Authors have declared no 'conflict of interest'.

Author details

Muhammad Ikram[1*], Ali Raza[2], Sarfraz Ali[2] and Salamat Ali[2]

1 Solar Cell Applications Research Lab, Department of Physics, Government College University Lahore, Lahore, Punjab, Pakistan

2 Department of Physics, Riphah Institute of Computing and Applied Sciences (RICAS), Riphah International University, Lahore, Pakistan

*Address all correspondence to: dr.muhammadikram@gcu.edu.pk

IntechOpen

© 2020 The Author(s). Licensee IntechOpen. This chapter is distributed under the terms of the Creative Commons Attribution License (http://creativecommons.org/licenses/by/3.0), which permits unrestricted use, distribution, and reproduction in any medium, provided the original work is properly cited. (cc) BY

References

[1] Q. Yang *et al.*, "Ultrathin graphene-based membrane with precise molecular sieving and ultrafast solvent permeation," *Nature Materials,* vol. 16, no. 12, pp. 1198-1202, 2017/12/01 2017.

[2] S. Goossens *et al.*, "Broadband image sensor array based on graphene–CMOS integration," *Nature Photonics,* vol. 11, no. 6, pp. 366-371, 2017/06/01 2017.

[3] H. Yan *et al.*, "Single-Atom Pd1/Graphene Catalyst Achieved by Atomic Layer Deposition: Remarkable Performance in Selective Hydrogenation of 1,3-Butadiene," *Journal of the American Chemical Society,* vol. 137, no. 33, pp. 10484-10487, 2015/08/26 2015.

[4] G. Qu *et al.*, "A Fiber Supercapacitor with High Energy Density Based on Hollow Graphene/Conducting Polymer Fiber Electrode," vol. 28, no. 19, pp. 3646-3652, 2016.

[5] A. Raza *et al.*, "Enhanced industrial dye degradation using Co doped in chemically exfoliated MoS2 nanosheets," *Applied Nanoscience,* vol. 10, no. 5, pp. 1535-1544, 2019.

[6] F. Calleja *et al.*, "Spatial variation of a giant spin–orbit effect induces electron confinement in graphene on Pb islands," *Nature Physics,* vol. 11, no. 1, pp. 43-47, 2015/01/01 2015.

[7] Y. Cao *et al.*, "Correlated insulator behaviour at half-filling in magic-angle graphene superlattices," *Nature,* vol. 556, no. 7699, pp. 80-84, 2018/04/01 2018.

[8] M. Ikram, M. I. Khan, A. Raza, M. Imran, A. Ul-Hamid, and S. Ali, "Outstanding performance of silver-decorated MoS2 nanopetals used as nanocatalyst for synthetic dye degradation," *Physica E: Low-dimensional Systems and Nanostructures,* vol. 124, 2020.

[9] S. Manzeli, D. Ovchinnikov, D. Pasquier, O. V. Yazyev, and A. Kis, "2D transition metal dichalcogenides," *Nature Reviews Materials,* vol. 2, no. 8, p. 17033, 2017/06/13 2017.

[10] U. Qumar *et al.*, "Synergistic effect of Bi-doped exfoliated MoS2 nanosheets on their bactericidal and dye degradation potential," *Dalton Trans,* vol. 49, no. 16, pp. 5362-5377, Apr 28 2020.

[11] J. A. Carrasco *et al.*, "Liquid phase exfoliation of carbonate-intercalated layered double hydroxides," *Chemical Communications,* 10.1039/C9CC00197B vol. 55, no. 23, pp. 3315-3318, 2019.

[12] W. Lei, V. N. Mochalin, D. Liu, S. Qin, Y. Gogotsi, and Y. Chen, "Boron nitride colloidal solutions, ultralight aerogels and freestanding membranes through one-step exfoliation and functionalization," *Nature Communications,* vol. 6, no. 1, p. 8849, 2015/11/27 2015.

[13] M. Ikram *et al.*, "Promising performance of chemically exfoliated Zr-doped MoS2 nanosheets for catalytic and antibacterial applications," *RSC Advances,* vol. 10, no. 35, pp. 20559-20571, 2020.

[14] A. Raza *et al.*, "A comparative study of dirac 2D materials, TMDCs and 2D insulators with regard to their structures and photocatalytic/sonophotocatalytic behavior," *Applied Nanoscience,* 2020.

[15] S. Yang, P. Zhang, A. S. Nia, and X. Feng, "Emerging 2D Materials Produced via Electrochemistry," vol. 32, no. 10, p. 1907857, 2020.

[16] A. Ambrosi and M. Pumera, "Exfoliation of layered materials using electrochemistry," *Chemical Society Reviews,* 10.1039/C7CS00811B vol. 47, no. 19, pp. 7213-7224, 2018.

[17] M. Ikram *et al.*, "Reduced graphene oxide nanosheets doped by Cu with highly efficient visible light photocatalytic behavior," *Journal of Alloys and Compounds,* vol. 837, 2020.

[18] H. Tao, Y. Zhang, Y. Gao, Z. Sun, C. Yan, and J. Texter, "Scalable exfoliation and dispersion of two-dimensional materials – an update," *Physical Chemistry Chemical Physics,* 10.1039/C6CP06813H vol. 19, no. 2, pp. 921-960, 2017.

[19] A. H. Khan *et al.*, "Two-Dimensional (2D) Nanomaterials towards Electrochemical Nanoarchitectonics in Energy-Related Applications," vol. 90, no. 6, pp. 627-648, 2017.

[20] A. M. Abdelkader, A. J. Cooper, R. A. W. Dryfe, and I. A. Kinloch, "How to get between the sheets: a review of recent works on the electrochemical exfoliation of graphene materials from bulk graphite," *Nanoscale,* 10.1039/C4NR06942K vol. 7, no. 16, pp. 6944-6956, 2015.

[21] M. Ikram *et al.*, "2D chemically exfoliated hexagonal boron nitride (hBN) nanosheets doped with Ni: synthesis, properties and catalytic application for the treatment of industrial wastewater," *Applied Nanoscience,* vol. 10, no. 9, pp. 3525-3528, 2020/09/01 2020.

[22] K. Parvez *et al.*, "Exfoliation of Graphite into Graphene in Aqueous Solutions of Inorganic Salts," *Journal of the American Chemical Society,* vol. 136, no. 16, pp. 6083-6091, 2014/04/23 2014.

[23] Y. Zhang, X. Zhang, Y. Ling, F. Li, A. M. Bond, and J. Zhang, "Controllable Synthesis of Few-Layer Bismuth Subcarbonate by Electrochemical Exfoliation for Enhanced CO2 Reduction Performance," vol. 57, no. 40, pp. 13283-13287, 2018.

[24] Y. Yang *et al.*, "Electrochemical exfoliation of graphene-like two-dimensional nanomaterials," vol. 11, no. 1, pp. 16-33, 2019.

[25] E. Hitz *et al.*, "Electrochemical Intercalation of Lithium Ions into NbSe2 Nanosheets," *ACS Applied Materials & Interfaces,* vol. 8, no. 18, pp. 11390-11395, 2016/05/11 2016.

[26] Y. Jung, Y. Zhou, and J. J. J. I. C. F. Cha, "Intercalation in two-dimensional transition metal chalcogenides," vol. 3, no. 4, pp. 452-463, 2016.

[27] G. Zhu *et al.*, "Tuning thermal conductivity in molybdenum disulfide by electrochemical intercalation," *Nature Communications,* vol. 7, no. 1, p. 13211, 2016/10/21 2016.

[28] Z. Lu, K. Jiang, G. Chen, H. Wang, and Y. Cui, "Lithium Electrochemical Tuning for Electrocatalysis," vol. 30, no. 48, p. 1800978, 2018.

[29] J. Xia *et al.*, "Phase evolution of lithium intercalation dynamics in 2H-MoS2," *Nanoscale,* 10.1039/C7NR02028G vol. 9, no. 22, pp. 7533-7540, 2017.

[30] S. Deng *et al.*, "Synergistic Doping and Intercalation: Realizing Deep Phase Modulation on MoS2 Arrays for High-Efficiency Hydrogen Evolution Reaction," vol. 58, no. 45, pp. 16289-16296, 2019.

[31] Z. Zeng *et al.*, "An Effective Method for the Fabrication of Few-Layer-Thick Inorganic Nanosheets," vol. 51, no. 36, pp. 9052-9056, 2012.

[32] J. Si *et al.*, "Scalable Production of Few-Layer Niobium Disulfide Nanosheets via Electrochemical Exfoliation for Energy-Efficient Hydrogen Evolution Reaction," *ACS Applied Materials & Interfaces,* vol. 11, no. 14, pp. 13205-13213, 2019/04/10 2019.

[33] Y. Yang *et al.*, "Electrochemical exfoliation of graphene-like two-dimensional nanomaterials," *Nanoscale,* vol. 11, no. 1, pp. 16-33, Dec 20 2018.

[34] D. Chen *et al.*, "Programmed electrochemical exfoliation of graphite to high quality graphene," *Chemical Communications,* 10.1039/C9CC00393B vol. 55, no. 23, pp. 3379-3382, 2019.

[35] E. Kovalska *et al.*, "Non-aqueous solution-processed phosphorene by controlled low-potential electrochemical exfoliation and thin film preparation," *Nanoscale,* 10.1039/C9NR10257D vol. 12, no. 4, pp. 2638-2647, 2020.

[36] J. M. Munuera, J. I. Paredes, S. Villar-Rodil, M. Ayán-Varela, A. Martínez-Alonso, and J. M. D. Tascón, "Electrolytic exfoliation of graphite in water with multifunctional electrolytes: en route towards high quality, oxide-free graphene flakes," *Nanoscale,* 10.1039/C5NR06882G vol. 8, no. 5, pp. 2982-2998, 2016.

[37] V. Nagyte *et al.*, "Raman Fingerprints of Graphene Produced by Anodic Electrochemical Exfoliation," *Nano Letters,* vol. 20, no. 5, pp. 3411-3419, 2020/05/13 2020.

[38] E. Kovalska, N. Antonatos, J. Luxa, and Z. Sofer, "“Top-down” Arsenene Production by Low-Potential Electrochemical Exfoliation," *Inorganic Chemistry,* vol. 59, no. 16, pp. 11259-11265, 2020/08/17 2020.

[39] K. Parvez, "Two-Dimensional Nanomaterials: Crystal Structure and Synthesis," in *Biomedical Applications of Graphene and 2D Nanomaterials*, 2019, pp. 1-25.

[40] Z. Zeng *et al.*, "Single-Layer Semiconducting Nanosheets: High-yield preparation and device fabrication," vol. 123, no. 47, pp. 11289-11293, 2011.

[41] Z. Zeng *et al.*, “Single-Layer Semiconducting Nanosheets: High-Yield Preparation and Device Fabrication,” vol. 50, no. 47, pp. 11093-11097, 2011.

[42] H. Wang *et al.*, "Electrochemical tuning of vertically aligned MoS_2 nanofilms and its application in improving hydrogen evolution reaction," vol. 110, no. 49, pp. 19701-19706, 2013.

[43] I. Jeon, B. Yoon, M. He, and T. M. Swager, "Hyperstage Graphite: Electrochemical Synthesis and Spontaneous Reactive Exfoliation," vol. 30, no. 3, p. 1704538, 2018.

[44] J. Wang, K. K. Manga, Q. Bao, and K. P. Loh, "High-Yield Synthesis of Few-Layer Graphene Flakes through Electrochemical Expansion of Graphite in Propylene Carbonate Electrolyte," *Journal of the American Chemical Society,* vol. 133, no. 23, pp. 8888-8891, 2011/06/15 2011.

[45] K. Parvez *et al.*, "Electrochemically Exfoliated Graphene as Solution-Processable, Highly Conductive Electrodes for Organic Electronics," *ACS Nano,* vol. 7, no. 4, pp. 3598-3606, 2013/04/23 2013.

[46] A. Ambrosi, Z. Sofer, and M. Pumera, "Electrochemical Exfoliation of Layered Black Phosphorus into Phosphorene," vol. 56, no. 35, pp. 10443-10445, 2017.

[47] Z. Huang, H. Hou, Y. Zhang, C. Wang, X. Qiu, and X. Ji, "Layer-Tunable Phosphorene Modulated by the Cation Insertion Rate as a Sodium-Storage Anode," vol. 29, no. 34, p. 1702372, 2017.

[48] L. Lu *et al.*, "Broadband Nonlinear Optical Response in Few-Layer Antimonene and Antimonene Quantum Dots: A Promising Optical Kerr Media with Enhanced Stability," vol. 5, no. 17, p. 1700301, 2017.

[49] X. Ren *et al.*, "Few-Layer Antimonene Nanosheet: A Metal-Free Bifunctional Electrocatalyst for Effective Water Splitting," *ACS Applied Energy Materials,* vol. 2, no. 7, pp. 4774-4781, 2019/07/22 2019.

[50] S. X. Leong, C. C. Mayorga-Martinez, X. Chia, J. Luxa, Z. Sofer, and M. Pumera, "2H → 1T Phase Change in Direct Synthesis of WS2 Nanosheets via Solution-Based Electrochemical Exfoliation and Their Catalytic Properties," *ACS Applied Materials & Interfaces,* vol. 9, no. 31, pp. 26350-26356, 2017/08/09 2017.

[51] A. Y. S. Eng, A. Ambrosi, Z. Sofer, P. Šimek, and M. Pumera, "Electrochemistry of Transition Metal Dichalcogenides: Strong Dependence on the Metal-to-Chalcogen Composition and Exfoliation Method," *ACS Nano,* vol. 8, no. 12, pp. 12185-12198, 2014/12/23 2014.

[52] Y. Zhang *et al.*, "Two-Dimensional Electrocatalysts for Efficient Reduction of Carbon Dioxide," vol. 13, no. 1, pp. 59-77, 2020.

[53] J. Huang *et al.*, "Electrochemical Exfoliation of Pillared-Layer Metal–Organic Framework to Boost the Oxygen Evolution Reaction," vol. 57, no. 17, pp. 4632-4636, 2018.

[54] S. Yang *et al.*, "Fluoride-Free Synthesis of Two-Dimensional Titanium Carbide (MXene) Using A Binary Aqueous System," vol. 57, no. 47, pp. 15491-15495, 2018.

[55] P. Yu *et al.*, "Mechanically-Assisted Electrochemical Production of Graphene Oxide," *Chemistry of Materials,* vol. 28, no. 22, pp. 8429-8438, 2016/11/22 2016.

[56] P. Yu, S. E. Lowe, G. P. Simon, and Y. L. Zhong, "Electrochemical exfoliation of graphite and production of functional graphene," *Current Opinion in Colloid & Interface Science,* vol. 20, no. 5, pp. 329-338, 2015/10/01/ 2015.

[57] A. J. Cooper, N. R. Wilson, I. A. Kinloch, and R. A. W. Dryfe, "Single stage electrochemical exfoliation method for the production of few-layer graphene via intercalation of tetraalkylammonium cations," *Carbon,* vol. 66, pp. 340-350, 2014/01/01/ 2014.

[58] S. García-Dalí *et al.*, "Aqueous Cathodic Exfoliation Strategy toward Solution-Processable and Phase-Preserved MoS2 Nanosheets for Energy Storage and Catalytic Applications," *ACS Applied Materials & Interfaces,* vol. 11, no. 40, pp. 36991-37003, 2019/10/09 2019.

[59] F. Luo *et al.*, "Ultrafast Cathodic Exfoliation of Few-Layer Black Phosphorus in Aqueous Solution," *ACS Applied Nano Materials,* vol. 2, no. 6, pp. 3793-3801, 2019/06/28 2019.

[60] J. M. Munuera, J. I. Paredes, S. Villar-Rodil, A. Martínez-Alonso, and J. M. D. Tascón, "A simple strategy to improve the yield of graphene nanosheets in the anodic exfoliation of graphite foil," *Carbon,* vol. 115, pp. 625-628, 2017/05/01/ 2017.

[61] F. Li *et al.*, "Advanced Composite 2D Energy Materials by Simultaneous Anodic and Cathodic Exfoliation," vol. 8, no. 12, p. 1702794, 2018.

[62] W. Wu, C. Zhang, L. Zhou, S. Hou, and L. Zhang, "High throughput synthesis of defect-rich MoS2 nanosheets via facile electrochemical exfoliation for fast high-performance lithium storage," *Journal of Colloid and Interface Science,* vol. 542, pp. 263-268, 2019/04/15/ 2019.

[63] M. S. Dresselhaus, "Fifty years in studying carbon-based materials," *Physica Scripta,* vol. T146, p. 014002, 2012/01/01 2012.

[64] A. Lerf, "Storylines in intercalation chemistry," *Dalton Transactions,* 10.1039/C4DT00203B vol. 43, no. 27, pp. 10276-10291, 2014.

[65] K. S. Novoselov *et al.*, "Electric Field Effect in Atomically Thin Carbon Films," *Science,* vol. 306, no. 5696, p. 666, 2004.

[66] C. Schafhaeutl, "Ueber die Verbindungen des Kohlenstoffes mit Silicium, Eisen und anderen Metallen, welche die verschiedenen Gallungen von Roheisen, Stahl und Schmiedeeisen bilden," vol. 21, no. 1, pp. 129-157, 1840.

[67] M. Noel and R. Santhanam, "Electrochemistry of graphite intercalation compounds," *Journal of Power Sources,* vol. 72, no. 1, pp. 53-65, 1998/03/30/ 1998.

[68] W. Rudorff and U. J. Z. A. A. C. Hofmann, "Über graphitsalze," vol. 238, pp. 1-50, 1938.

[69] M. Inagaki, "Applications of graphite intercalation compounds," *Journal of Materials Research,* vol. 4, no. 6, pp. 1560-1568, 1989.

[70] G.-A. Nazri and G. Pistoia, *Lithium batteries: science and technology*. Springer Science & Business Media, 2008.

[71] M. Noel, R. Santhanam, and M. Francisca Flora, "Effect of polypyrrole film on the stability and electrochemical activity of fluoride based graphite intercalation compounds in HF media," *Journal of Applied Electrochemistry,* vol. 24, no. 5, pp. 455-459, 1994/05/01 1994.

[72] M. Noel, R. Santhanam, and M. J. B. o. e. Franciscaflora, "Studies of fluoride intercalation in graphite electrodes-the effect of HF concentration on competitive anodic processes," vol. 12, no. 07, pp. 421-425, 1996.

[73] H. Takenaka, M. Kawaguchi, M. Lerner, and N. Bartlett, "Synthesis and characterization of graphite fluorides by electrochemical fluorination in aqueous and anhydrous hydrogen fluoride," *Journal of the Chemical Society, Chemical Communications,* 10.1039/C39870001431 no. 19, pp. 1431-1432, 1987.

[74] H. Y. Zhang, W. C. Shen, Z. D. Wang, and F. Zhang, "Formation of iron chloride-graphite intercalation compounds in propylene carbonate by electrolysis," *Carbon,* vol. 35, no. 2, pp. 285-290, 1997/01/01/ 1997.

[75] T. Ohzuku, Y. Iwakoshi, and K. Sawai, "Formation of Lithium-Graphite Intercalation Compounds in Nonaqueous Electrolytes and Their Application as a Negative Electrode for a Lithium Ion (Shuttlecock) Cell," *Journal of The Electrochemical Society,* vol. 140, no. 9, pp. 2490-2498, 1993/09/01 1993.

[76] Y. Nishi, "Lithium ion secondary batteries; past 10 years and the future," *Journal of Power Sources,* vol. 100, no. 1, pp. 101-106, 2001/11/30/ 2001.

[77] N. Iwashita and M. Inagaki, "Discharge performance on the electrode of ternary intercalation compounds of graphite with metal hydroxides," *Electrochimica Acta,* vol. 36, no. 3, pp. 591-595, 1991/01/01/ 1991.

[78] Y. L. Zhong and T. M. Swager, "Enhanced Electrochemical Expansion of Graphite for in Situ Electrochemical Functionalization," *Journal of the American Chemical Society,* vol. 134, no. 43, pp. 17896-17899, 2012/10/31 2012.

[79] J. Xu *et al.*, "Recent Progress in Graphite Intercalation Compounds for Rechargeable Metal (Li, Na, K, Al)-Ion Batteries," vol. 4, no. 10, p. 1700146, 2017.

[80] J. Liu *et al.*, "A green approach to the synthesis of high-quality graphene oxide flakes via electrochemical exfoliation of pencil core," *RSC Advances,* 10.1039/

C3RA41366G vol. 3, no. 29, pp. 11745-11750, 2013.

[81] C.-Y. Su, A.-Y. Lu, Y. Xu, F.-R. Chen, A. N. Khlobystov, and L.-J. Li, "High-Quality Thin Graphene Films from Fast Electrochemical Exfoliation," *ACS Nano,* vol. 5, no. 3, pp. 2332-2339, 2011/03/22 2011.

[82] M. Mao, M. Wang, J. Hu, G. Lei, S. Chen, and H. Liu, "Simultaneous electrochemical synthesis of few-layer graphene flakes on both electrodes in protic ionic liquids," *Chemical Communications,* 10.1039/C3CC41909F vol. 49, no. 46, pp. 5301-5303, 2013.

[83] A. M. Abdelkader, I. A. Kinloch, and R. A. W. Dryfe, "Continuous Electrochemical Exfoliation of Micrometer-Sized Graphene Using Synergistic Ion Intercalations and Organic Solvents," *ACS Applied Materials & Interfaces,* vol. 6, no. 3, pp. 1632-1639, 2014/02/12 2014.

[84] J. Liu, M. Notarianni, G. Will, V. T. Tiong, H. Wang, and N. Motta, "Electrochemically Exfoliated Graphene for Electrode Films: Effect of Graphene Flake Thickness on the Sheet Resistance and Capacitive Properties," *Langmuir,* vol. 29, no. 43, pp. 13307-13314, 2013/10/29 2013.

[85] N. E. Sorokina, N. V. Maksimova, and V. V. Avdeev, "Anodic Oxidation of Graphite in 10 to 98% HNO3," *Inorganic Materials,* vol. 37, no. 4, pp. 360-365, 2001/04/01 2001.

[86] Y. Wang, C. C. Mayorga-Martinez, X. Chia, Z. Sofer, and M. Pumera, "Nonconductive layered hexagonal boron nitride exfoliation by bipolar electrochemistry," *Nanoscale,* 10.1039/C8NR00082D vol. 10, no. 15, pp. 7298-7303, 2018.

[87] H. Xiao *et al.*, "Electrochemical cathode exfoliation of bulky black phosphorus into few-layer phosphorene nanosheets," *Electrochemistry Communications,* vol. 89, pp. 10-13, 2018/04/01/ 2018.

[88] S. Yang *et al.*, "A Delamination Strategy for Thinly Layered Defect-Free High-Mobility Black Phosphorus Flakes," vol. 57, no. 17, pp. 4677-4681, 2018.

[89] T. C. Achee *et al.*, "High-yield scalable graphene nanosheet production from compressed graphite using electrochemical exfoliation," *Scientific Reports,* vol. 8, no. 1, p. 14525, 2018/09/28 2018.

[90] A. J. Jacobson and L. F. Nazar, "Intercalation Chemistry," in *Encyclopedia of Inorganic and Bioinorganic Chemistry*, 2011.

[91] Y. Fang *et al.*, "Janus electrochemical exfoliation of two-dimensional materials," *Journal of Materials Chemistry A,* 10.1039/C9TA10487A vol. 7, no. 45, pp. 25691-25711, 2019.

[92] K. S. Rao, J. Senthilnathan, Y.-F. Liu, and M. Yoshimura, "Role of Peroxide Ions in Formation of Graphene Nanosheets by Electrochemical Exfoliation of Graphite," *Scientific Reports,* vol. 4, no. 1, p. 4237, 2014/02/28 2014.

[93] A. Ambrosi *et al.*, "Graphene and its electrochemistry – an update," *Chemical Society Reviews,* 10.1039/C6CS00136J vol. 45, no. 9, pp. 2458-2493, 2016.

[94] L. Hu *et al.*, "Direct anodic exfoliation of graphite onto high-density aligned graphene for large capacity supercapacitors," *Nano Energy,* vol. 34, pp. 515-523, 2017/04/01/ 2017.

[95] C.-Y. Lee, D. R. G. Mitchell, P. Molino, A. Fahy, and G. G. Wallace, "Tunable solution-processable anodic exfoliated graphene," *Applied Materials Today,* vol. 15, pp. 290-296, 2019/06/01/ 2019.

[96] Z. Y. Xia *et al.*, "Synergic Exfoliation of Graphene with Organic Molecules and Inorganic Ions for the Electrochemical Production of Flexible Electrodes," vol. 79, no. 3, pp. 439-446, 2014.

[97] S. Yang *et al.*, "Organic Radical-Assisted Electrochemical Exfoliation for the Scalable Production of High-Quality Graphene," *Journal of the American Chemical Society,* vol. 137, no. 43, pp. 13927-13932, 2015/11/04 2015.

[98] J. M. Munuera *et al.*, "Electrochemical Exfoliation of Graphite in Aqueous Sodium Halide Electrolytes toward Low Oxygen Content Graphene for Energy and Environmental Applications," *ACS Applied Materials & Interfaces,* vol. 9, no. 28, pp. 24085-24099, 2017/07/19 2017.

[99] Z. Liu *et al.*, "Water-Dispersed High-Quality Graphene: A Green Solution for Efficient Energy Storage Applications," *ACS Nano,* vol. 13, no. 8, pp. 9431-9441, 2019/08/27 2019.

[100] A. Ejigu *et al.*, "On the Role of Transition Metal Salts During Electrochemical Exfoliation of Graphite: Antioxidants or Metal Oxide Decorators for Energy Storage Applications," vol. 28, no. 48, p. 1804357, 2018.

[101] C.-H. Chen, S.-W. Yang, M.-C. Chuang, W.-Y. Woon, and C.-Y. Su, "Towards the continuous production of high crystallinity graphene via electrochemical exfoliation with molecular in situ encapsulation," *Nanoscale,* 10.1039/C5NR03669K vol. 7, no. 37, pp. 15362-15373, 2015.

[102] S. Yang *et al.*, "Ultrafast Delamination of Graphite into High-Quality Graphene Using Alternating Currents," vol. 56, no. 23, pp. 6669-6675, 2017.

[103] L. Li *et al.*, "Regulation of radicals from electrochemical exfoliation of a double-graphite electrode to fabricate high-quality graphene," *Journal of Materials Chemistry C,* 10.1039/C8TC01565A vol. 6, no. 23, pp. 6257-6263, 2018.

[104] A. Ambrosi and M. Pumera, "Electrochemically Exfoliated Graphene and Graphene Oxide for Energy Storage and Electrochemistry Applications," vol. 22, no. 1, pp. 153-159, 2016.

[105] J. Cao *et al.*, "Two-Step Electrochemical Intercalation and Oxidation of Graphite for the Mass Production of Graphene Oxide," *Journal of the American Chemical Society,* vol. 139, no. 48, pp. 17446-17456, 2017/12/06 2017.

[106] S. Pei, Q. Wei, K. Huang, H.-M. Cheng, and W. Ren, "Green synthesis of graphene oxide by seconds timescale water electrolytic oxidation," *Nature Communications,* vol. 9, no. 1, p. 145, 2018/01/10 2018.

[107] M. Xu *et al.*, "Lithium-assisted exfoliation of pristine graphite for few-layer graphene nanosheets," *Nano Research,* vol. 8, no. 3, pp. 801-807, 2015/03/01 2015.

[108] L. Liu *et al.*, "High-Yield Chemical Vapor Deposition Growth of High-Quality Large-Area AB-Stacked Bilayer Graphene," *ACS Nano,* vol. 6, no. 9, pp. 8241-8249, 2012/09/25 2012.

[109] H. Huang *et al.*, "Highly efficient electrolytic exfoliation of graphite into graphene sheets based on Li ions intercalation–expansion–microexplosion mechanism," *Journal of Materials Chemistry,* 10.1039/C2JM00092J vol. 22, no. 21, pp. 10452-10456, 2012.

[110] C.-J. Shih, S. Lin, M. S. Strano, and D. Blankschtein, "Understanding the Stabilization of Liquid-Phase-Exfoliated Graphene in Polar Solvents: Molecular Dynamics Simulations and

Kinetic Theory of Colloid Aggregation," *Journal of the American Chemical Society,* vol. 132, no. 41, pp. 14638-14648, 2010/10/20 2010.

[111] M. Zhou, J. Tang, Q. Cheng, G. Xu, P. Cui, and L.-C. Qin, "Few-layer graphene obtained by electrochemical exfoliation of graphite cathode," *Chemical Physics Letters,* vol. 572, pp. 61-65, 2013/05/30/ 2013.

[112] A. J. Cooper, M. Velický, I. A. Kinloch, and R. A. W. Dryfe, "On the controlled electrochemical preparation of R4N+ graphite intercalation compounds and their host structural deformation effects," *Journal of Electroanalytical Chemistry,* vol. 730, pp. 34-40, 2014/09/15/ 2014.

[113] Y. Yang, F. Lu, Z. Zhou, W. Song, Q. Chen, and X. Ji, "Electrochemically cathodic exfoliation of graphene sheets in room temperature ionic liquids N-butyl, methylpyrrolidinium bis(trifluoromethylsulfonyl)imide and their electrochemical properties," *Electrochimica Acta,* vol. 113, pp. 9-16, 2013/12/15/ 2013.

[114] X. Lu and C. Zhao, "Controlled electrochemical intercalation, exfoliation and in situ nitrogen doping of graphite in nitrate-based protic ionic liquids," *Physical Chemistry Chemical Physics,* 10.1039/C3CP53917B vol. 15, no. 46, pp. 20005-20009, 2013.

[115] D. Wei *et al.*, "Graphene from electrochemical exfoliation and its direct applications in enhanced energy storage devices," *Chemical Communications,* 10.1039/C2CC16859F vol. 48, no. 9, pp. 1239-1241, 2012.

Chapter 6

Carbon Nanotubes

Muhammad Ikram, Ali Raza, Atif Shahbaz, Haleema Ijaz, Sarfraz Ali, Ali Haider, Muhammad Tayyab Hussain, Junaid Haider, Arslan Ahmed Rafi and Salamat Ali

Abstract

Carbon nanotubes (CNTs) are referred to as carbon nano-architecture allotropes, with wrapped graphene sheets forming a cylindrical structure. CNTs are either developed by metals or narrow-band semiconductors with rolling graphene sheets in various ways. Researchers have dedicated a great deal of attention to understanding the fascinating properties of CNTs over the years, and possess certain peculiar properties, such as a high degree of stiffness, a wide ratio of length to diameter, and remarkable toughness, and are employed in a number of applications. These properties can be enhanced by scheming the diameter, nature of walls, chirality, length of CNTs which is rolled up, and depending on the synthesis process. This chapter extensively covers the various properties of CNTs and how it influences to desired applications and also explains numerous methods of synthesis and processing of CNTs with advantages and some drawbacks.

Keywords: 2D materials, graphene oxide, carbon nanotube, arc discharge, laser ablation

1. Introduction

In recent decades, formation of nanowires and nanotubes has an attractive literature which emphasizes towards material growth. Amongst numerous materials containing organic and inorganic, nanotubes show versatile properties due to promising candidates such like carbon nanotubes, contributing a great part in potential applications relevant to disciplinary medicinal chemistry [1, 2]. Foundation of fullerenes [3] was extracted from carbon nanotubes (CNTs) that explore fabrication on a macroscopic level, thereby exhibiting continuous evolution [4]. The cylindrical shape of CNT is caused by rolling up of graphitic sheets; length is measured in micrometer scale while maximum diameter is taken as 100 nm. CNT also appears in bundle shape to form prominently complex nature structure [5]. Hexagon rings are in arranged form on which metallic nature or semiconducting behavior of CNT is evaluated. CNT belongs to the properties towards robust applications like fillers; bio-sensors are amongst nanotechnological pillars in exciting fields [6, 7]. However, some limitations such as insolubility and non-manipulation in solvents play role for creating hindrance to CNT use as solute in organic solvents as well as aqueous media. Dispersion of CNT may be carried out through sonication; however, precipitation is also occurred caused by the interruption of the process followed by the mechanism. Moreover, numerous studies also showed that CNT might react with a variety of chemical compounds [8–17].

Innovative nanodevices are greatly desired in research work and it may be met only by CNT best fabricating processing that is obtained by the synthesis of complex nature composites [18–20]. Furthermore, CNTs become highly reliable when chemical reactions are carried out to incorporate them in soluble activities into different systems such as organic or inorganic and biological accordingly. Thus, CNTs solubility approach in chemical reactions opens new routes for introducing promising materials [21, 22]. Unidirectional CNT structures may be prepared by modified approaches and their structural study is done by following group study containing three categories, first is that various chemical groups are incorporated on the surface of CNTs via covalent bonding, secondly non-covalent wrapping of functional groups and thirdly endohedral fulfillment of cavity. Many citations in this study have been appreciated due to which it is rapidly increasing by worth in literature, while this review presents a limited approach providing useful information in all citations followed in this study [23–26]. It has been systematically studied that CNTs may be prepared by employing synthesis methods containing arc discharge approach or chemical-vapor-deposition and laser-ablation technique [27, 28].

In arc discharge approach temperature is kept greater than 3000°C. This temperature is indispensable to evaporate carbon atoms to form a plasma state, in this way CNTs are shaped as single-walled as well as multi-walled structures. In this process, catalytic agent may or may not be involved during the formation of multi-walled carbon nanotubes (MWCNTs). However, inclusion of catalytic agent is mandatory to create individual single-walled carbon nanotubes (SWCNTs). The catalytic agents like Cobalt, Nickel, and Iron may be used as mandatory steps to complete the reactions reasonably [29–32]. In chemical-vapor-deposition (CVD) approach methane, ethylene, etc. are incorporated as hydrocarbon sources necessary to carry out reactions successfully. As far as laser-ablation approach is concerned, evaporation process of graphite occurs in a furnace at a temperature of 1200°C. Moreover, graphite appears as dominant material to produce species with converting ratio at maximum level. Moreover, biomaterial targets are achieved depending on degree of purity level, that is why macroscopic approach is carried out for the improved quality of carbon materials owing to achieve some characteristics like length and alignment [33]. It has been reported that MWCNTs were collected first time by Iijima (by employing arc-discharge approach), and this approach is too old that was adopted for carbon fibers synthesis [34, 35]. Subsequently, an in-situ emulsion of polymerization was presented by Khan et al. [36] in 2016 to synthesize carbon nanotubes (CNTs) in the form of composites, which was completed by employing a colloidal system to fabricate nanostructured brush.

2. Classification of carbon nanotubes

Nanotubes may be categorized into SWCNTs as well as MWCNTs (see **Figure 1**). A comparison between both SWCNT and MWCNT is demonstrated in **Table 1** [38, 41].

3. Structure and morphology

SWCNT comprised of carbon atoms from graphene sheet containing benzene rings in hexagonal shape as illustrated in **Figure 2a**. Cylindrical graphene sheets comprising honeycomb lattice are visualized in single-atomic graphitic-layer of crystalline nature, while MWCNT is in stacked form of graphene sheets that are rolled up into cylinders having same centers. The composition of nanotube

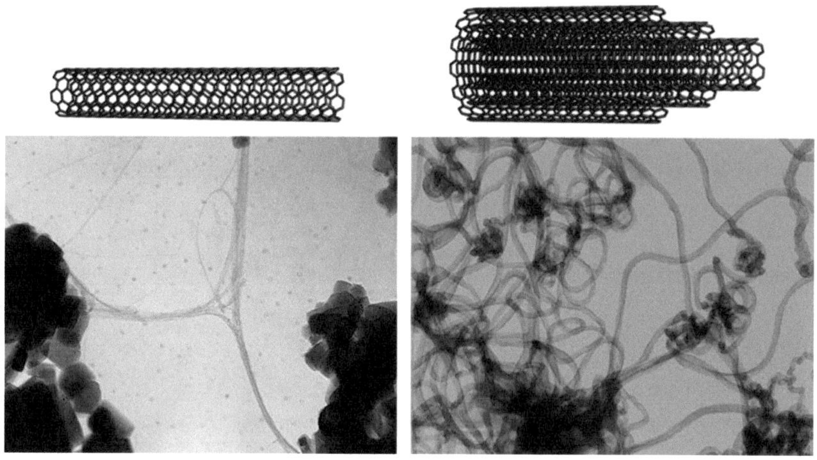

Figure 1.
Molecular representations of SWCNT (top left) and MWCNT (top right) with typical transmission electron micrographs below [37].

SWCNT	MWCNT
Single-layer of graphene.	Multiple layers of graphene
Catalyst is required for synthesis.	Can be produced without catalyst.
Bulk synthesis is difficult as it requires proper control over growth and atmospheric condition.	Bulk synthesis is easy.
Not fully dispersed, and form bundled bundled formation.	Homogeneously dispersed with no apparent structures.
Resistivity usually in the range of $10^{-4} - 10^{-3}$ Ω.m	Resistivity usually in the range of $1.8 \times 10^{-5} - 6.1 \times 10^{-5}$ Ω·m
Purity is poor. Typical SWCNT content in as-prepared samples by chemical vapor deposition (CVD) method is about 30–50 wt%. However high purity up to 80% has been reported by using arc discharge synthesis method.	Purity is high. Typical MWCNT content in as-prepared samples by CVD method is about 35–90 wt%.
A chance of defect is more during functionalization.	A chance of defect is less especially when synthesized by an arc-discharged method.
Characterization and evaluation are easy.	It has a very complex structure
It can be easily twisted and are more pliable.	It cannot be easily twisted.

Table 1.
Comparison between SWCNT and MWCNT [38–40].

molecules contains a million atoms having length of tens micrometers and diameters are comparable with 0.7 nm value [41]. SWCNTs containing 10 atoms often lie along the circumference of tube-like structure with one-atom-thick thickness. A length to diameter ratio of carbon nanotubes is measured about 1000 (large aspect ratio), giving rise to be considered as unidirectional structures [43]. MWCNTs structure is formed by various single-walled tubes that are stacked in concentric cylinders inside each other. The MWCNTs are identified as nanostructures showing the outer diameter is (15 nm or less) while structures having a diameter more than 15 nm are considered as nanofibers, not nanotubes. CNTs are different from carbon-fibers owing to not a single (molecule) yet strand layers sheets of graphitic nature [43–46].

Depending upon the two aforesaid basic structures, carbon nanotubes may be categorized into three varieties as an armchair, zigzag, and chiral carbon nanotubes.

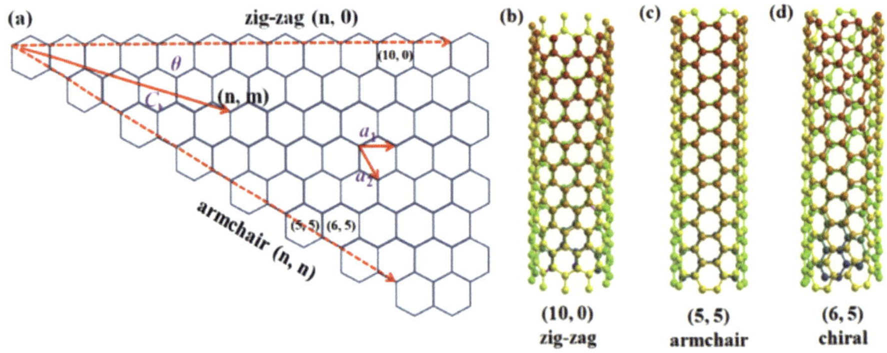

Figure 2.
(a) Unrolled single-layer graphene sheet showing the geometry of the SWCNT, (b-d) Examples of the three types of nanotube sidewall; zigzag, armchair, and chiral (A color version of this figure can be viewed online) [42].

The structure of a variety of carbon nanotubes depends on the formation of rolled up graphitic cylinders during synthesis process. The main focus is selection of rolling-axis relative to hexagonal graphitic network of sheets as well as radius of closing cylindrical network of nanotubes that are raised in various types of SWCNTs. In this structure, chiral vector contains *n* and *m* indices corresponding to two unit vectors directing along two-axis in graphene crystal lattice structure. In case of m = 0 zigzag-type nanotube, but when n = m the armchair nanotube is obtained while other configurations are attributed to chiral type nanotubes accordingly. In addition, SWCNTs with armchair structure, zigzag, and chiral structures have been illustrated in **Figure 2b–d**. Moreover, a further detailed structure may be visualized in literature reviews [7, 43, 47, 48].

4. Properties

Mechanical properties may drastically be raised, caused by the electrostatic forces between sp^2 carbon–carbon-bonds. Previously no material has been yet found to display the collective mechanical, electronic, and thermal properties up till now. Densities of materials have been observed below 1.3 g/cm^3 value (one-sixth stainless steel). Young's moduli measured material stiffness that was greater than 1 TPa and is considered approximately 5x higher than that of stainless steel [49, 50]. However, uniqueness of materials still depends upon strength that makes them apart from others. Furthermore, carbon nanotubes are those materials that showed the strongest stiffness in the history of mankind. The tensile strength of carbon nanotubes measured so far is up to 63 GPa that is considered about 50 times greater than that of stainless steel [51]. However, carbon nanotubes that are identified as the weakest one show only several GPa strength [52]. As far as chemical, environmental stability, thermal conductivity etc. are compared to diamond. Owing to such attractive properties along with lightness of carbon nanotubes opens new routes towards variety of applications particularly in the field of aerospace [40, 53–55].

Carbon nanotubes highly exhibit electronic properties as compared to other materials. On comparing with copper carbon nanotubes show an extraordinary electrical conductivity. The most notable fact here is metallic as well as semiconducting nature of carbon nanotubes. The rolled-up structure comes forward to break up symmetric shape of the planar system. In this way, different directions are observed attributing to hexagonal lattice of carbon material and also axial direction is disturbed. Axial direction and unit vectors describe hexagonal lattice, therefore, depending on electrical properties carbon nanotubes may have nature of metal or semiconducting material. Amongst other nanotubes, semiconducting nanotubes may have band gap inversely with diameter. Band gap range was found between (1.8–0.18 eV) relative to small diameter tubes as well as very wide SWCNT respectively [56, 57]. Consequently, various nanotubes may belong to higher conductivity as compared to copper metal, while some others relative to silicon have a more conducting nature. There is a still promising interest in fabrication of nanoscale electronic devices by active use of nanotubes. Various areas of technology need carbon nanotubes to prepare advanced materials. Thus carbon nanotubes are already frequently used in those areas of research. Some outcomes of nanotubes are flat-panel displays, fuel cells, scanning probe microscopes, and sensing fabricated devices [58].

4.1 Optical properties

Electronic properties owing to SWNTs have been theoretically studied in early decades. SWNTs may be predicted metallic or semiconductors based on parameters that are followed in structure formation of the nanotubes [35]. As far as metallic and semiconducting nature of nanotubes is concerned, one third belongs to metallic whereas two-third relates to semiconducting nanotubes concerning selected indices (n, m). The aforesaid model is identified as π tight-binding model related to zone-folding scheme. Tight-binding data is based on (σ and π) bands that produce the curvature of σ and π bands. This bending behavior indicates a very small gap lying between metallic and semiconducting nanotubes [59, 60].

4.2 Electrical properties

Electrical properties of carbon nanotubes show electrical transport impact that becomes an interesting area of various possible applications attributing to fabricate electronic devices at nanoscale basis. Nanotubes are classified as one-dimensional conductor owing to which attractive microscopic phenomena are observed at low temperatures. Phenomena are likewise single-electron charging, superconductivity, and resonant tunneling. On the other hand, high temperature based tunneling conductance expresses power-law suppression that is evaluated as a function of (temperature and bias voltage) consistent with one-dimensional Luttinger liquid. Scattering mechanism is raised by optical or zone-boundary-phonons in metal-like nanotubes. Scattering along with coherent-backscattering phenomena has resulted in the form of low-temperature phenomena. Probe measurements were two-type as well as four-type in transport experiments performed with respect to MWNTs [61], isolated SWNTs, and SWNT bundles respectively [62, 63].

Initially, electrical resistance was measured towards unique MWNT below $T = 20$ mK, Langer et al. determined [61], whereas magnetic field shows a logarithmic conductance trend at declining temperature whereas saturation level was identified at the temperature below $T \sim 0.3$ K. However, when magnetic field impact was measured and found perpendicular towards tube axis, at that time magneto-resistance measurements were also observed. Furthermore, temperature effect on conductance in magnetic field was also observed that was found inconsistent with two-dimensional weak-localization.

4.3 Vibrational properties

Atomic-vibrations into carbon nanotubes were successfully evaluated by employing force-constant models (zone-folding-approximation) [64], also for concrete structure of nanotubes [65], ranging (tight-binding-models) [66–69] and finally, ab-initio models were also observed [70]. To measure vibrational eigenfrequencies, experiments were performed by using light resonant Raman scattering in case of laser-light-energy when energy measurements are very close to available electronic transitions. Resonance limitations are entirely different for all types of nanotubes; therefore Raman spectroscopy presents results to display various nanotubes structures that exist in the nanotube specimen. Currently, Raman spectroscopy measured parallel polarized light relevant to MWNTs [71], SWNTs [72, 73] and cross-polarized-light on isolated SWNTs [74].

4.4 Thermal properties

Phonons were used to measure specific heat as well as thermal conductivity of carbon nanotube systems. When temperature was kept low enough, acoustic phonons were observed indicating dominant role of phonon contribution in the nanotube systems. Linear specific heat measurements and thermal conductivity yield at or above 1 K but below room-temperature [75, 76], whereas 0.62 T specific heat identifies temperature at or below 1 K [77]. Linear temperature was evaluated depending on linear k-vector and modes of vibration of acoustic phonons such as longitudinal and twist like vibrations [78]. Transverse acoustic phonons are relative to specific heat exhibits dependence behavior attributing to specific heat at or below 1 K along with quadratic k-vector trend [79]. Thermoelectric measurement power (TEMP) for nanotube systems presents active and direct information about carrier types along with conductivity mechanisms [80–83].

5. Synthesis of CNTs

High-quality carbon nanotubes are considered to be superior quality materials and proved to be main pillar towards promising and versatile applications, various synthesis routes are employed to achieve feasible application of CNTs as described in **Figure 3**. Superior quality indicates that density of structural defects is significantly less over length scale between 1 to 10 microns along tube-axes. Carbon nanotubes synthesis is rapidly increasing in research field but still, challenges are prevailing. Those challenges are required to resolve with respect to synthesis of CNT. The main challenges are of four types regarding nanotube synthesis [84]. First is mass-production scale, containing low-cost based synthesis with large-scale synthetic routes to produce high-quality SWCNTs nanotubes. Second is a selective production scale that raises control over structural defects and changes electronic properties relevant to produced nanotubes. Third is Organization level regarding control over location along with specific orientation towards produced nanotubes on specific substrate. Fourth is mechanism level that presents all procedures followed during growth of nanotubes in synthetic processes. But growth mechanism is considered still controversial because alternative mechanisms may be employed during fabrication of CNTs [27, 85, 86].

Different techniques have been systematically employed to develop and produce SWNTs as well as MWNTs showing various structural and morphological characters in laboratory quantities. Methods commonly followed are three in number to synthesize CNTs, first is arc discharge [87, 88], second is laser ablation [66, 89] and third is chemical vapor deposition [67–69, 90, 91]. Catalysts are considered basic elements that are selected as source of carbon towards nanotubes formation, having sufficient energy. A significant feature of all methods followed for CNTs

Figure 3.
Currently used methods for CNTs synthesis [84].

fabrication is to enhance energy for carbon source producing fragments of carbon atoms that may recombine to yield SWNTs or MWNTs. The main goal is source of energy that is electricity and heat from an arc discharge and CVD respectively or high-intensity-light for laser ablation.

5.1 Arc discharge and laser vaporization

Amongst various methods that were allowed regarding SWNTs synthesis, arc-discharge or laser-ablation methods contributed relatively on large-scale basis (**Figure 4**). Subsequently, carbons atoms in a gaseous state are condensed caused by evaporation process of solid-state carbon atoms [92]. While growing single-wall-nanotubes (SWCNTs) in arc-discharge system, metallic catalyst is mandatorily required to incorporate for speed-up desired chemical reactions [93]. On the other hand, superior-quality (SWCNTs) are successfully fabricated (1–10 g scale) by using a laser oven approach [94]. Besides aforementioned method wave, CO_2-laser system was also employed regarding industrial-scale production of SWCNTs [95]. However, costly equipment as well as high energy consumption requirement makes them unfavorable approaches towards production of nanotube materials. Through employing arc ablation or laser methods only powder type specimens of carbon materials into bundle-shape form are controllably produced. The most common characteristic relevant to arc-discharge and laser-ablation approaches indicates higher energy need to induce carbon atoms to rearrange forming CNTs. Favorable temperature is prominently 3000 °C (or higher value) that is considered more beneficial for fine crystallization growth of CNTs at this level since products are obtained with attractive graphite-alignment. Moreover basic needs of the systems such as vacuum-conditions, repeated graphite-target substitution create barriers towards production of CNTs on an industrial scale [96].

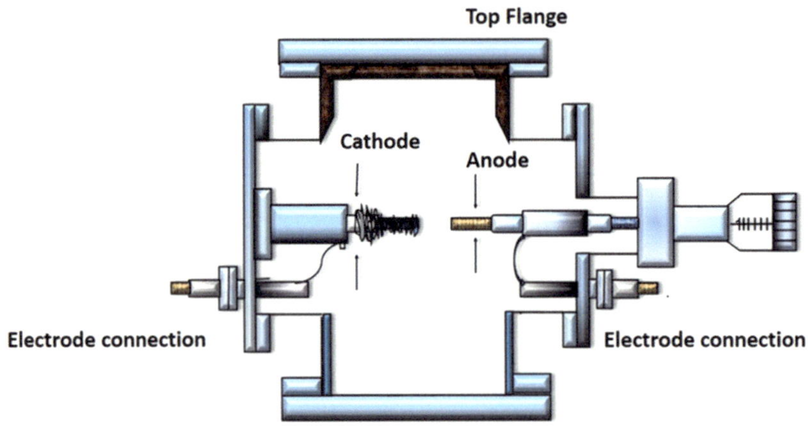

Figure 4.
Schematic diagram showing the Arc discharge method [40].

5.2 Chemical vapor deposition (CVD)

CVD approach presents carbon compounds decomposition in gaseous state where metallic nanoparticles are used as catalysts resulting in nucleation sites available for initial growth of carbon nanotubes (CNTs). Main drawback found in previous both methods was lack of large-scale fabrication of carbon materials, but CVD approach has presented preferred route towards carbon nanotubes production at large scale [97–99]. In this work, carbon is extracted from hydrocarbon source or some other carbon generating source. These chemical reactions are only successfully performed by using catalysts at or below 1200 °C temperature. Resultantly, CNT structure involved parameters like wall number, length, alignment, and diameter that have proven controllable CVD process. In addition, CVD approach has greater scope and advantages over other methods showing mild operation with low cost and selective process. The previous twelve years period describe that various approaches have presented promising industrial-scale synthesis of carbon nanotubes. All approaches indicate that CVD methods are main pillars of large scale production of nanotubes. Among various methods, main approaches are five in numbers that have proven to be successful large-scale yield [100].

i. Methane(CH_4) chemical vapor deposition (CVD) approach has been reported in 1998 showing bulky synthesis of SWCNTs by employing CVD method directly from CH_4 at temperature level (900 °C) [101, 102]. But Su *et al.* [103] successfully enhanced yield efficiency of CVD method by using aluminum oxide (Al_2O_3) aerogels. These aerogels were impregnated with a catalyst containing Fe/Mo nanoparticles.

ii. High-Pressure carbon monooxide (HPCO) approach presented catalytic-decomposition of CO through CO (a carbon source) at high pressure towards SWCNTs fabrication [104]. Aforesaid catalysts were used in decomposition process in the form of a gas phase state collected from organometallic-catalyst that was used as a precursor.

iii. In CO-CVD approach CO gas plays role to feed gas. On comparing with previously prepared samples with same methane catalyst, amorphous carbon rate was drastically reduced. In addition, the use of Co-Mo catalyst was considered an additional advancement in previously employed CO-CVD approach [105]. This approach incorporates Co/Mo bimetallic catalyst along with fluidized CVD-reactor during production of SWCNTs at large scale. The main benefit that arises from controllable use of fluidized CVD reactors was that they stop continuous addition as well as removal of solid-like particles during operation without stopping reactor work.

iv. Alcoholic CVD approach has been presented by Maruyama *et al*. in 2002 report [106] by producing superior quality SWCNTs with aless amount of amorphous carbon. Alcohols were used in methanol or ethanol from carbon source. Resultantly, formation of OH radical during synthesis process played a vital role for maximum removal of amorphous carbon from efficient growth of pure SWCNTs leaving them advanced material in a research field.

v. Currently Plasma Enhanced Chemical Vapor Deposition (PECVD) approach has been used widely towards fabrication of MWCNTs and SWCNTs carboneous compounds [107–109]. Moreover reactive species present in plasma-system may seriously affect growth of carbon nanotubes with small diameter, thereby creating implications for diameter control along with selective etching attributing to metallic SWCNTs growth.

5.3 Laser ablation method

Both laser ablation and arc discharge approaches have the same principles with similar mechanisms. However, they are not similar with respect to energy sources that are adopted to complete reactions. A laser is main source of desired energy for laser ablation method and **Figure 5** is showing schematic experimental setup. The schematic structure contains quartz tube with graphitic block. Graphite block is heated at 1200°C temperature by using high-power-laser whereas metal particles are incorporated catalysts [110]. Argon gas is controllably used in stream form during reaction process. Graphite lying in quartz is systematically vaporized by functioning of laser. Argon present in chamber removes vapors of carbon by condensation process towards downstream cooler quartz-walls. Condensation process is completed in the presence of both SWCNTs and metallic nanoparticles (see **Figure 5**). Literature reveals that laser power may

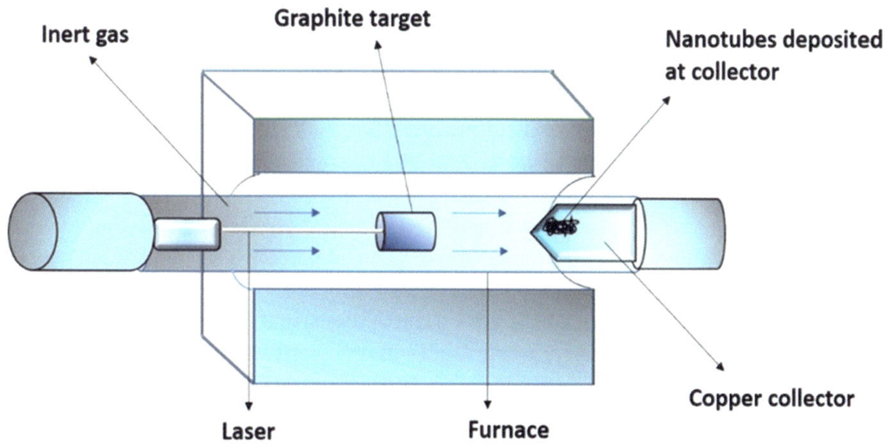

Figure 5.
Schematic structure showing the laser ablation method.

strongly affect CNTs diameter. On increasing laser pulse power rate a very thin diameter carbon nanotube is collected [40]. On other hand, some other reports give more information in favor of laser pulses that they behave like great potential, owing to be capable to provide SWCNTs in large amount [89, 111]. Reports indicate pure and superior quality production of SWCNTs in this case. Curved graphene sheets are observed showing position of carbon atoms in condense phase state caused by set up created by metal-catalyst to fabricate condensed carbon nanotubes.

In this case, carbon atoms rearrange them for formation of ring shape and in this way, electronegative properties become dominant to play role in preventing open edge from sealing [110]. Furthermore, there are main benefits relative to this method that indicate metallic impurities less in amount but high in yield owing to vaporization tendency creating at tube end of metallic atoms at closing position. However main drawback of technique is observed with respect to synthesis aspect of nanotubes that they are not regularly straight rather indicate degree of branching to some extent. In addition this technique involves high-quality graphitic rods with high-power laser rate. However, in this case, CNTs are produced but not greater than arc-discharge technique. Carbon-2019 for "PEER REVIEW" describes high-power-laser when metal particles are incorporated as mandatory catalysts in reaction process [110]. Argon gaseous stream is continuously used during reaction mechanism. Graphitic quartz is passed through vaporization process using a laser, argon media captures carbon vapors that result in condensed downstream towards cooler-walls of quartz but still SWCNTs with metallic-particles are located in condensation process. Laser power may also clearly affect CNTs diameter. Furthermore, diameter becomes comparatively narrowed on increasing laser pulse rate [30]. Other studies reported that ultrafast laser pulses are of great potential, and are capable to produce larger quantities of SWCNTs [112]. SWCNTs collected by this technique are observed owing to high-purity and superior-quality in nature. Location sites where carbons atoms initiate condensation process may set up curved shape graphene sheet along with metal-catalyst atoms. In this way condensed nanotubes are properly obtained showing peculiar properties. Moreover, carbon atoms merge to form specific rings, thereby raising electronegative properties relative to metallic atoms that become capable to prevent open-edge from closing [113].

The main benefit belongs to followed method, in this case, metallic impurities are observed relatively less in amount but with high yield that is caused by vapors formation tendency belonging to metallic atoms from tube end when closed once in a time. However main drawback relative to this technique indicates irregularity in straight shape for synthesized nanotubes whereas degree of branching occurs to some extent. Furthermore, pure graphitic rods are involved in this procedure along with high laser power rate. Resultantly production of CNTs was not in great amount as compared to arc- discharge method.

6. Conclusions

Carbon nanotubes have the ability to be more investigated, and it is possible to drive further advancements by using CNTs in different fields. The findings obtained in the synthesis, functionalization, and structure of CNTs have contributed significantly to promising developments in various fields. However, further perfections in synthesis protocols are needed to obtain highly durable CNTs for preferred applications. For an instant, catalyst size is directly influenced on diameter of CNT during CVD reaction. So, further analysis should also be undertaken to discover more effective methods of processing precisely uniform-sized catalyst particles in order to ensure the production desired diameter of SWCNTs; but CNTs are costly than other carbon nanomaterials. Efforts should be proceeded to look for modern, cost-effective, and plentiful carbon sources, so that cost of CNTs can be lowered to an acceptable amount.

Author details

Muhammad Ikram[1*], Ali Raza[2], Atif Shahbaz[1], Haleema Ijaz[3], Sarfraz Ali[2], Ali Haider[4], Muhammad Tayyab Hussain[2], Junaid Haider[5], Arslan Ahmed Rafi[2] and Salamat Ali[2]

1 Department of Physics, Government College University Lahore, Punjab, Pakistan

2 Department of Physics, Riphah Institute of Computing and Applied Sciences (RICAS), Riphah International University, Lahore, Pakistan

3 Department of Allied and Health Sciences, Superior College University Campus, Lahore, Pakistan

4 Department of Clinical Medicine and Surgery, University of Veterinary and Animal Sciences, Lahore, Punjab, Pakistan

5 Tianjin Institute of Industrial Biotechnology, Chinese Academy of Sciences, Tianjin, China

*Address all correspondence to: dr.muhammadikram@gcu.edu.pk

IntechOpen

© 2020 The Author(s). Licensee IntechOpen. This chapter is distributed under the terms of the Creative Commons Attribution License (http://creativecommons.org/licenses/by/3.0), which permits unrestricted use, distribution, and reproduction in any medium, provided the original work is properly cited.

References

[1] Kinloch IA, Suhr J, Lou J, Young RJ, Ajayan PM. Composites with carbon nanotubes and graphene: An outlook. Science. 2018;**362**(6414):547

[2] He X et al. Carbon nanotubes as emerging quantum-light sources. In: *Nature Materials,* vol. 17, no. 8, pp. 663-670, 2018/08/01. 2018

[3] Gatti T, Menna E, Meneghetti M, Maggini M, Petrozza A, Lamberti F. The Renaissance of fullerenes with perovskite solar cells. In: *Nano Energy,* vol. 41, pp. 84-100, 2017/11/01/. 2017

[4] Tasis D, Tagmatarchis N, Bianco A, Prato M. Chemistry of Carbon Nanotubes. In: *Chemical Reviews,* vol. 106, no. 3, pp. 1105-1136, 2006/03/01. 2006

[5] Takakura A et al. Strength of carbon nanotubes depends on their chemical structures. In: *Nature Communications,* vol. 10, no. 1, p. 3040, 2019/07/10. 2019

[6] S. A. Miners, G. A. Rance, and A. N. Khlobystov, "Chemical reactions confined within carbon nanotubes," *Chemical Society Reviews,* 10.1039/C6CS00090H vol. 45, no. 17, pp. 4727-4746, 2016.

[7] Zaporotskova IV, Boroznina NP, Parkhomenko YN, Kozhitov LV. Carbon nanotubes: Sensor properties. A review. In: *Modern Electronic Materials,* vol. 2, no. 4, pp. 95-105, 2016/12/01/. 2016

[8] Thostenson ET, Ren Z, Chou T-W. Advances in the science and technology of carbon nanotubes and their composites: a review. In: *Composites Science and Technology,* vol. 61, no. 13, pp. 1899-1912, 2001/10/01/. 2001

[9] Lin D, Xing B. Adsorption of Phenolic Compounds by Carbon Nanotubes: Role of Aromaticity and Substitution of Hydroxyl Groups. In: *Environmental Science & Technology,* vol. 42, no. 19, pp. 7254-7259, 2008/10/01. 2008

[10] Chen J, Chen W, Zhu D. Adsorption of Nonionic Aromatic Compounds to Single-Walled Carbon Nanotubes: Effects of Aqueous Solution Chemistry. In: *Environmental Science & Technology,* vol. 42, no. 19, pp. 7225-7230, 2008/10/01. 2008

[11] Pignatello JJ, Mitch WA, Xu W. Activity and Reactivity of Pyrogenic Carbonaceous Matter toward Organic Compounds. In: *Environmental Science & Technology,* vol. 51, no. 16, pp. 8893-8908, 2017/08/15. 2017

[12] Lee H, Lee H-J, Jeong J, Lee J, Park N-B, Lee C. Activation of persulfates by carbon nanotubes: Oxidation of organic compounds by nonradical mechanism. In: *Chemical Engineering Journal,* vol. 266, pp. 28-33, 2015/04/15/. 2015

[13] Liu SF, Moh LCH, Swager TM. Single-Walled Carbon Nanotube–Metalloporphyrin Chemiresistive Gas Sensor Arrays for Volatile Organic Compounds. In: *Chemistry of Materials,* vol. 27, no. 10, pp. 3560-3563, 2015/05/26. 2015

[14] Vázquez E, Prato M. Carbon Nanotubes and Microwaves: Interactions, Responses, and Applications. In: *ACS Nano,* vol. 3, no. 12, pp. 3819-3824, 2009/12/22. 2009

[15] Ma PC, Tang BZ, Kim J-K. Effect of CNT decoration with silver nanoparticles on electrical conductivity of CNT-polymer composites. In: *Carbon,* vol. 46, no. 11, pp. 1497-1505, 2008/09/01/. 2008

[16] D. Janas, K. Z. Milowska, P. D. Bristowe, and K. K. K. Koziol, "Improving the electrical properties of carbon nanotubes with interhalogen

compounds," *Nanoscale,* 10.1039/C7NR00224F vol. 9, no. 9, pp. 3212-3221, 2017.

[17] Raza A et al. Enhanced industrial dye degradation using Co doped in chemically exfoliated MoS2 nanosheets. Applied Nanoscience. 2019;**10**(5):1535-1544

[18] Q. Shi et al., "A Vision-Based Automated Manipulation System for the Pick-Up of Carbon Nanotubes," IEEE/ASME Transactions on Mechatronics, vol. 22, no. 2, pp. 845-854, 2017.

[19] Lin C-H, Fu H-C, Lien D-H, Hsu C-Y, He J-H. Self-powered nanodevices for fast UV detection and energy harvesting using core-shell nanowire geometry. In: *Nano Energy,* vol. 51, pp. 294-299, 2018/09/01/. 2018

[20] W. D. Yang, Y. D. Li, and X. Wang, "Scale-Dependent Dynamic-Pull-In of Functionally Graded Carbon Nanotubes Reinforced Nanodevice with Piezoelectric Layer," vol. 30, no. 3, p. 04016096, 2017.

[21] Scarselli M, Castrucci P, De Crescenzi M. Electronic and optoelectronic nano-devices based on carbon nanotubes. In: *Journal of Physics: Condensed Matter,* vol. 24, no. 31, p. 313202, 2012/07/11. 2012

[22] Tanaka H, Yajima T, Matsumoto T. Y. Otsuka, and T. Ogawa, "Porphyrin Molecular Nanodevices Wired Using Single-Walled Carbon Nanotubes,". 2006;**18**(11):1411-1415

[23] Q. W. Li *et al.*, "Structure-Dependent Electrical Properties of Carbon Nanotube Fibers," vol. 19, no. 20, pp. 3358-3363, 2007.

[24] Chen C-M, Dai Y-M, Huang JG, Jehng J-M. Intermetallic catalyst for carbon nanotubes (CNTs) growth by thermal chemical vapor deposition method. In: *Carbon,* vol. 44, no. 9, pp. 1808-1820, 2006/08/01/. 2006

[25] Chernyak SA et al. Co catalysts supported on oxidized CNTs: Evolution of structure during preparation, reduction and catalytic test in Fischer-Tropsch synthesis. In: *Applied Catalysis A: General,* vol. 523, pp. 221-229, 2016/08/05/. 2016

[26] Raza A et al. A comparative study of dirac 2D materials, TMDCs and 2D insulators with regard to their structures and photocatalytic/sonophotocatalytic behavior. Applied Nanoscience. 2020

[27] Shah KA, Tali BA. Synthesis of carbon nanotubes by catalytic chemical vapour deposition: A review on carbon sources, catalysts and substrates. In: *Materials Science in Semiconductor Processing,* vol. 41, pp. 67-82, 2016/01/01/. 2016

[28] Jourdain V, Bichara C. Current understanding of the growth of carbon nanotubes in catalytic chemical vapour deposition. In: *Carbon,* vol. 58, pp. 2-39, 2013/07/01/. 2013

[29] Anzar N, Hasan R, Tyagi M, Yadav N, Narang J. Carbon nanotube - A review on Synthesis, Properties and plethora of applications in the field of biomedical science. In: *Sensors International,* vol. 1, p. 100003, 2020/01/01/. 2020

[30] Yang Y, Zhang H, Yan Y. Synthesis of CNTs on stainless steel microfibrous composite by CVD: Effect of synthesis condition on carbon nanotube growth and structure. In: *Composites Part B: Engineering,* vol. 160, pp. 369-383, 2019/03/01/. 2019

[31] D. Wang *et al.*, "In Situ Synthesis of CNTs/Cu Nanocomposites and the Electronic Transport Properties," vol. 256, no. 8, p. 1800557, 2019.

[32] Ikram M, Khan MI, Raza A, Imran M, Ul-Hamid A, Ali S. Outstanding performance of silver-decorated MoS2 nanopetals used as nanocatalyst for synthetic dye degradation. *Physica E: Low-dimensional Systems and Nanostructures,* vol. 2020:124

[33] He H, Pham-Huy LA, Dramou P, Xiao D, Zuo P, Pham-Huy C. Carbon Nanotubes: Applications in Pharmacy and Medicine. In: *BioMed Research International,* vol. 2013, p. 578290, 2013/09/30. 2013

[34] Baughman RH, Zakhidov AA, de Heer WA. Carbon Nanotubes--the Route Toward Applications. Science. 2002;**297**(5582):787

[35] Popov VN. Carbon nanotubes: properties and application. In: *Materials Science and Engineering: R: Reports,* vol. 43, no. 3, pp. 61-102, 2004/01/15/. 2004

[36] Khan MU, Reddy KR, Snguanwongchai T, Haque E, Gomes VG. Polymer brush synthesis on surface modified carbon nanotubes via in situ emulsion polymerization. In: *Colloid and Polymer Science,* vol. 294, no. 10, pp. 1599-1610, 2016/10/01. 2016

[37] Donaldson K et al. Carbon Nanotubes: A Review of Their Properties in Relation to Pulmonary Toxicology and Workplace Safety. Toxicological Sciences. 2006;**92**(1):5-22

[38] Saifuddin N, Raziah AZ, Junizah AR. Carbon Nanotubes: A Review on Structure and Their Interaction with Proteins. Journal of Chemistry. 2013;**2013**:1-18

[39] J. Abrahamson, P. G. Wiles, and B. L. J. C. Rhoades, "Structure of carbon fibres found on carbon arc anodes," vol. 37, no. 11, pp. 1873-1874, 1999.

[40] Rahman G et al. An Overview of the Recent Progress in the Synthesis and Applications of Carbon Nanotubes. *C,* vol. 2019;**1**:5

[41] Meyyappan M, Delzeit L, Cassell A, Hash D. Carbon nanotube growth by PECVD: a review. In: *Plasma Sources Science and Technology,* vol. 12, no. 2, pp. 205-216, 2003/04/02. 2003

[42] Zhang F, Hou P-X, Liu C, Cheng H-M. Epitaxial growth of single-wall carbon nanotubes. In: *Carbon,* vol. 102, pp. 181-197, 2016/06/01/. 2016

[43] Dresselhaus MS. G. Dresselhaus, and A. Jorio, "UNUSUAL PROPERTIES AND STRUCTURE OF CARBON NANOTUBES,". 2004;**34**(1):247-278

[44] Lin T, Bajpai V. T. Ji, and L. Dai, "Chemistry of Carbon Nanotubes %J Australian Journal of Chemistry,". 2003;**56**(7):635-651

[45] Flebus B, MacDonald AH. Electronic structure of carbon nanotubes on graphene substrates. In: *Physical Review Research,* vol. 2, no. 2, p. 022041, 05/15/. 2020

[46] Ikram M et al. Reduced graphene oxide nanosheets doped by Cu with highly efficient visible light photocatalytic behavior. *Journal of Alloys and Compounds,* vol. 2020:837

[47] M. Terrones, "Science and Technology of the Twenty-First Century: Synthesis, Properties, and Applications of Carbon Nanotubes," vol. 33, no. 1, pp. 419-501, 2003.

[48] Imani Yengejeh S, Kazemi SA, Öchsner A. Carbon nanotubes as reinforcement in composites: A review of the analytical, numerical and experimental approaches. In: *Computational Materials Science,* vol. 136, pp. 85-101, 2017/08/01/. 2017

[49] Yu M-F, Files BS, Arepalli S, Ruoff RS. Tensile Loading of Ropes

of Single Wall Carbon Nanotubes and their Mechanical Properties. In: *Physical Review Letters,* vol. 84, no. 24, pp. 5552-5555, 06/12/. 2000

[50] Farrera C, Torres Andón F, Feliu N. Carbon Nanotubes as Optical Sensors in Biomedicine. In: *ACS Nano,* vol. 11, no. 11, pp. 10637-10643, 2017/11/28. 2017

[51] Mousavi-Khoshdel SM, Jahanbakhsh-bonab P, Targholi E. Structural, electronic properties, and quantum capacitance of B, N and P-doped armchair carbon nanotubes. In: *Physics Letters A,* vol. 380, no. 41, pp. 3378-3383, 2016/10/07/. 2016

[52] Xie S, Li W, Pan Z, Chang B, Sun L. Mechanical and physical properties on carbon nanotube. In: *Journal of Physics and Chemistry of Solids,* vol. 61, no. 7, pp. 1153-1158, 2000/07/01/. 2000

[53] Li W-k, Shi Y-p. Recent advances and applications of carbon nanotubes based composites in magnetic solid-phase extraction. In: *TrAC Trends in Analytical Chemistry,* vol. 118, pp. 652-665, 2019/09/01/. 2019

[54] Wang R, Xie L, Hameed S, Wang C, Ying Y. Mechanisms and applications of carbon nanotubes in terahertz devices: A review. In: *Carbon,* vol. 132, pp. 42-58, 2018/06/01/. 2018

[55] Poudel YR, Li W. Synthesis, properties, and applications of carbon nanotubes filled with foreign materials: a review. In: *Materials Today Physics,* vol. 7, pp. 7-34, 2018/12/01/. 2018

[56] Elliott JA, Sandler JKW, Windle AH, Young RJ, Shaffer MSP. Collapse of Single-Wall Carbon Nanotubes is Diameter Dependent. In: *Physical Review Letters,* vol. 92, no. 9, p. 095501, 03/02/. 2004

[57] Zawadzka A, Płóciennik P, Korcala A, Szroeder P. Optical properties of chiral single-walled carbon nanotubes thin films. In: *Optical Materials,* vol. 96, p. 109295, 2019/10/01/. 2019

[58] Sanli A, Benchirouf A, Müller C, Kanoun O. Piezoresistive performance characterization of strain sensitive multi-walled carbon nanotube-epoxy nanocomposites. In: *Sensors and Actuators A: Physical,* vol. 254, pp. 61-68, 2017/02/01/. 2017

[59] Charlier JC, Lambin P. Electronic structure of carbon nanotubes with chiral symmetry. In: *Physical Review B,* vol. 57, no. 24, pp. R15037-R15039, 06/15/. 1998

[60] Arnold MS, Green AA, Hulvat JF, Stupp SI, Hersam MC. Sorting carbon nanotubes by electronic structure using density differentiation. In: *Nature Nanotechnology,* vol. 1, no. 1, pp. 60-65, 2006/10/01. 2006

[61] Langer L et al. Quantum Transport in a Multiwalled Carbon Nanotube. In: *Physical Review Letters,* vol. 76, no. 3, pp. 479-482, 01/15/. 1996

[62] Bockrath M et al. Single-Electron Transport in Ropes of Carbon Nanotubes. Science. 1997;**275**(5308):1922

[63] Fischer JE et al. Metallic resistivity in crystalline ropes of single-wall carbon nanotubes. In: *Physical Review B,* vol. 55, no. 8, pp. R4921-R4924, 02/15/. 1997

[64] Eklund PC, Holden JM, Jishi RA. Vibrational modes of carbon nanotubes; Spectroscopy and theory. In: *Carbon,* vol. 33, no. 7, pp. 959-972, 1995/01/01/. 1995

[65] Popov VN, Van Doren VE, Balkanski M. Lattice dynamics of single-walled carbon nanotubes. In: *Physical Review B,* vol. 59, no. 13, pp. 8355-8358, 04/01/. 1999

[66] Vander Wal RL, Berger GM, Ticich TM. Carbon nanotube synthesis in a flame using laser ablation for in situ catalyst generation. In: *Applied Physics A,* vol. 77, no. 7, pp. 885-889, 2003/12/01. 2003

[67] Li Y-L, Kinloch IA, Windle AH. Direct Spinning of Carbon Nanotube Fibers from Chemical Vapor Deposition Synthesis. Science. 2004;**304**(5668):276

[68] See CH, Harris AT. A Review of Carbon Nanotube Synthesis via Fluidized-Bed Chemical Vapor Deposition. In: *Industrial & Engineering Chemistry Research,* vol. 46, no. 4, pp. 997-1012, 2007/02/01. 2007

[69] Jung M, Yong Eun K, Lee J-K, Baik Y-J, Lee K-R, Wan Park J. Growth of carbon nanotubes by chemical vapor deposition. In: *Diamond and Related Materials,* vol. 10, no. 3, pp. 1235-1240, 2001/03/01/. 2001

[70] Dubay O, Kresse G. Accurate density functional calculations for the phonon dispersion relations of graphite layer and carbon nanotubes. In: *Physical Review B,* vol. 67, no. 3, p. 035401, 01/06/. 2003

[71] Rao AM et al. Polarized Raman Study of Aligned Multiwalled Carbon Nanotubes. In: *Physical Review Letters,* vol. 84, no. 8, pp. 1820-1823, 02/21/. 2000

[72] Jorio A et al. Polarized Raman Study of Single-Wall Semiconducting Carbon Nanotubes. In: *Physical Review Letters,* vol. 85, no. 12, pp. 2617-2620, 09/18/. 2000

[73] Azoulay J, Débarre A, Richard A, Tchénio P, Bandow S, Iijima S. Polarised Raman spectroscopy on a single class of single-wall nanotubes by nano surface-enhanced scattering. In: *Chemical Physics Letters,* vol. 331, no. 5, pp. 347-353, 2000/12/08/. 2000

[74] Jorio A et al. Resonance Raman Spectra of Carbon Nanotubes by Cross-Polarized Light. In: *Physical Review Letters,* vol. 90, no. 10, p. 107403, 03/14/. 2003

[75] Yi W, Lu L, Dian-lin Z, Pan ZW, Xie SS. Linear specific heat of carbon nanotubes. In: *Physical Review B,* vol. 59, no. 14, pp. R9015-R9018, 04/01/. 1999

[76] Balandin AA. Thermal properties of graphene and nanostructured carbon materials. In: *Nature Materials,* vol. 10, no. 8, pp. 569-581, 2011/08/01. 2011

[77] Lasjaunias JC, Biljaković K, Benes Z, Fischer JE, Monceau P. Low-temperature specific heat of single-wall carbon nanotubes. In: *Physical Review B,* vol. 65, no. 11, p. 113409, 02/22/. 2002

[78] Kuan H-C, Ma C-CM, Chang W-P, Yuen S-M, Wu H-H, Lee T-M. Synthesis, thermal, mechanical and rheological properties of multiwall carbon nanotube/waterborne polyurethane nanocomposite. In: *Composites Science and Technology,* vol. 65, no. 11, pp. 1703-1710, 2005/09/01/. 2005

[79] Popov VN. Low-temperature specific heat of nanotube systems. In: *Physical Review B,* vol. 66, no. 15, p. 153408, 10/29/. 2002

[80] Ma PC, Kim J-K, Tang BZ. Effects of silane functionalization on the properties of carbon nanotube/epoxy nanocomposites. In: *Composites Science and Technology,* vol. 67, no. 14, pp. 2965-2972, 2007/11/01/. 2007

[81] Wang J, Xie H, Xin Z. Thermal properties of paraffin based composites containing multi-walled carbon nanotubes. In: *Thermochimica Acta,* vol. 488, no. 1, pp. 39-42, 2009/05/05/. 2009

[82] Chen X et al. Mechanical and thermal properties of epoxy nanocomposites reinforced with amino-functionalized multi-walled carbon

nanotubes. In: *Materials Science and Engineering: A,* vol. 492, no. 1, pp. 236-242, 2008/09/25/. 2008

[83] Song YS, Youn JR. Influence of dispersion states of carbon nanotubes on physical properties of epoxy nanocomposites. In: *Carbon,* vol. 43, no. 7, pp. 1378-1385, 2005/06/01/. 2005

[84] J. Prasek *et al.*, "Methods for carbon nanotubes synthesis—review," *Journal of Materials Chemistry,* 10.1039/C1JM12254A vol. 21, no. 40, pp. 15872-15884, 2011.

[85] Hong EH, Lee K-H. S. H. Oh, and C.-G. Park, "Synthesis of Carbon Nanotubes Using Microwave Radiation,". 2003;**13**(12):961-966

[86] Yadav MD, Dasgupta K, Patwardhan AW, Joshi JB. High Performance Fibers from Carbon Nanotubes: Synthesis, Characterization, and Applications in Composites—A Review. In: *Industrial & Engineering Chemistry Research,* vol. 56, no. 44, pp. 12407-12437, 2017/11/08. 2017

[87] Vekselman V, Feurer M, Huang T, Stratton B, Raitses Y. Complex structure of the carbon arc discharge for synthesis of nanotubes. In: *Plasma Sources Science and Technology,* vol. 26, no. 6, p. 065019, 2017/06/06. 2017

[88] Arora N, Sharma NN. Arc discharge synthesis of carbon nanotubes: Comprehensive review. In: *Diamond and Related Materials,* vol. 50, pp. 135-150, 2014/11/01/. 2014

[89] J. Chrzanowska *et al.*, "Synthesis of carbon nanotubes by the laser ablation method: Effect of laser wavelength," vol. 252, no. 8, pp. 1860-1867, 2015.

[90] M. Ikram, A. Raza, M. Imran, A. Ul-Hamid, A. Shahbaz, and S. Ali, "Hydrothermal Synthesis of Silver Decorated Reduced Graphene Oxide (rGO) Nanoflakes with Effective Photocatalytic Activity for Wastewater Treatment," *Nanoscale Res Lett,* vol. 15, no. 1, p. 95, Apr 28 2020.

[91] Ikram M et al. 2D chemically exfoliated hexagonal boron nitride (hBN) nanosheets doped with Ni: synthesis, properties and catalytic application for the treatment of industrial wastewater. In: *Applied Nanoscience,* vol. 10, no. 9, pp. 3525-3528, 2020/09/01. 2020

[92] Tahriri M et al. Graphene and its derivatives: Opportunities and challenges in dentistry. In: *Materials Science and Engineering: C,* vol. 102, pp. 171-185, 2019/09/01/. 2019

[93] Zhonghai LZ, JI DTANG, Chang LIU, Huiming CHENG. A Review on the Controlled Growth of Single-Wall Carbon Nanotubes from Metal Catalysts. In: vol. 54, no. 11, pp. 1665-1682, 2018-11-11. 2018

[94] Thess A et al. Crystalline Ropes of Metallic Carbon Nanotubes. Science. 1996;**273**(5274):483

[95] Muñoz E et al. Single-walled carbon nanotubes produced by cw CO2-laser ablation: study of parameters important for their formation. In: *Applied Physics A,* vol. 70, no. 2, pp. 145-151, 2000/02/01. 2000

[96] Malekimoghadam R, Rafiee R. 3 - Carbon Nanotubes Processing. In: Polymers CN-R, Rafiee R, editors. Elsevier. 2018. pp. 41-59

[97] Dai H. Carbon nanotubes: opportunities and challenges. In: *Surface Science,* vol. 500, no. 1, pp. 218-241, 2002/03/10/. 2002

[98] Krasnovskii AN, Kishchuk PS. Effect of catalyst mass on CVD synthesis of carbon nanotubes. In: *Russian Journal of Applied Chemistry,* vol. 90, no. 5, pp. 721-725, 2017/05/01. 2017

[99] A. Hussain *et al.*, "Floating catalyst CVD synthesis of single walled carbon nanotubes from ethylene for high performance transparent electrodes," *Nanoscale*, 10.1039/C8NR00716K vol. 10, no. 20, pp. 9752-9759, 2018.

[100] Muñoz-Sandoval E, Fajardo-Díaz JL, Sánchez-Salas R, Cortés-López AJ, López-Urías F. Two Sprayer CVD Synthesis of Nitrogen-doped Carbon Sponge-type Nanomaterials. In: *Scientific Reports*, vol. 8, no. 1, p. 2983, 2018/02/14. 2018

[101] Kong J, Cassell AM, Dai H. Chemical vapor deposition of methane for single-walled carbon nanotubes. In: *Chemical Physics Letters*, vol. 292, no. 4, pp. 567-574, 1998/08/14/. 1998

[102] Cassell AM, Raymakers JA, Kong J, Dai H. Large Scale CVD Synthesis of Single-Walled Carbon Nanotubes. In: *The Journal of Physical Chemistry B*, vol. 103, no. 31, pp. 6484-6492, 1999/08/01. 1999

[103] Su M, Zheng B, Liu J. A scalable CVD method for the synthesis of single-walled carbon nanotubes with high catalyst productivity. In: *Chemical Physics Letters*, vol. 322, no. 5, pp. 321-326, 2000/05/26/. 2000

[104] Nikolaev P et al. Gas-phase catalytic growth of single-walled carbon nanotubes from carbon monoxide. In: *Chemical Physics Letters*, vol. 313, no. 1, pp. 91-97, 1999/11/05/. 1999

[105] Kitiyanan B, Alvarez WE, Harwell JH, Resasco DE. Controlled production of single-wall carbon nanotubes by catalytic decomposition of CO on bimetallic Co–Mo catalysts. In: *Chemical Physics Letters*, vol. 317, no. 3, pp. 497-503, 2000/02/04/. 2000

[106] Maruyama S, Kojima R, Miyauchi Y, Chiashi S, Kohno M. Low-temperature synthesis of high-purity single-walled carbon nanotubes from alcohol. In: *Chemical Physics Letters*, vol. 360, no. 3, pp. 229-234, 2002/07/10/. 2002

[107] Li Y et al. Preferential Growth of Semiconducting Single-Walled Carbon Nanotubes by a Plasma Enhanced CVD Method. In: *Nano Letters*, vol. 4, no. 2, pp. 317-321, 2004/02/01. 2004

[108] Zhong G, Iwasaki T, Honda K, Furukawa Y, Ohdomari I, Kawarada H. Low Temperature Synthesis of Extremely Dense and Vertically Aligned Single-Walled Carbon Nanotubes. In: *Japanese Journal of Applied Physics*, vol. 44, no. 4A, pp. 1558-1561, 2005/04/08. 2005

[109] Zhang G et al. Ultra-high-yield growth of vertical single-walled carbon nanotubes: Hidden roles of hydrogen and oxygen. Proceedings of the National Academy of Sciences of the United States of America. 2005;**102**(45):16141

[110] Eatemadi A et al. Carbon nanotubes: properties, synthesis, purification, and medical applications. In: *Nanoscale Research Letters*, vol. 9, no. 1, p. 393, 2014/08/13. 2014

[111] Mwafy EA, Mostafa AM. Multi walled carbon nanotube decorated cadmium oxide nanoparticles via pulsed laser ablation in liquid media. In: *Optics & Laser Technology*, vol. 111, pp. 249-254, 2019/04/01/. 2019

[112] Sakakibara Y, Tatsuura S, Kataura H, Tokumoto M, Achiba Y. Near-Infrared Saturable Absorption of Single-Wall Carbon Nanotubes Prepared by Laser Ablation Method. In: *Japanese Journal of Applied Physics*, vol. 42, no. Part 2, No. 5A, pp. L494-L496, 2003/05/01. 2003

[113] Scott CD, Arepalli S, Nikolaev P, Smalley RE. Growth mechanisms for single-wall carbon nanotubes in a laser-ablation process. In: *Applied Physics A*, vol. 72, no. 5, pp. 573-580, 2001/05/01. 2001